ENVIRONMENTAL SCIENCE, ENGINEERING AND TECHNOLOGY

TROUBLED WATERS: ASIAN CARP AND THE GREAT LAKES

Environmental Science, Engineering and Technology

Additional books in this series can be found on Nova's website under the Series tab.

Additional E-books in this series can be found on Nova's website under the E-books tab.

Water Resource Planning, Development and Management

Additional books in this series can be found on Nova's website under the Series tab.

Additional E-books in this series can be found on Nova's website under the E-books tab.

ENVIRONMENTAL SCIENCE, ENGINEERING AND TECHNOLOGY

TROUBLED WATERS: ASIAN CARP AND THE GREAT LAKES

DEBORAH M. MAEDA
EDITOR

Nova Science Publishers, Inc.
New York

For permission to use material from this book please contact us:
Telephone 631-231-7269; Fax 631-231-8175
Web Site: http://www.novapublishers.com

Additional color graphics may be available in the e-book version of this book.

LIBRARY OF CONGRESS CATALOGING-IN-PUBLICATION DATA

Asian carp and the Great Lakes.
Troubled waters : Asian carp and the Great Lakes / editor, Deborah M. Maeda.
p. cm.
Consists of hearing before the Subcommittee on Water Resources and Environment of the House Committee on Transportation and Infrastructure, February 9, 2010, with a Congressional Research Service report, Asian carp and the Great Lakes region, dated June 25, 2010.
Hearing originally published: Asian carp and the Great Lakes. Washington : U.S.G.P.O., 2010.
Includes index.
ISBN 978-1-61761-766-9 (hardcover)
1. Carp--Control--Great Lakes (North America) 2. Introduced fishes--Control--Great Lakes (North America) 3. Nonindigenous pests--Control--Great Lakes (North America) I. Maeda, Deborah M. II. United States. Congress. House. Committee on Transportation and Infrastructure. Subcommittee on Water Resources and Environment. III. Library of Congress. Congressional Research Service IV. Title.
QL638.C94A77 2010
333.95'6482140977--dc22
2010041329

Published by Nova Science Publishers, Inc. † New York

CONTENTS

PREFACE

Asian carp were first introduced into the southern United States in the early 1970s to remove algae from farmed catfish ponds. Flooding in the early 1900s caused many of the catfish farm ponds to overflow, and Asian carp were released into local waterways in the Mississippi River basin. The carp have since migrated northward up the Mississippi River, becoming the most abundant species in some areas of the River. Asian carp are of particular concern to the Great Lakes region because of the potential harm they could cause to the native ecosystem. Asian carp currently present in the Mississippi River physically lack a stomach, so they must continually feed on aquatic vegetation. Native fish species have difficulty competing with Asian carp because of their rapid consumption of resources. This book provides an overview and background of the threat of Asian carp migrating into the Great Lakes and what can be done to avoid this threat.

Chapter 1- Four species of non-indigenous Asian carp are expanding their range in U.S. waterways, resulting in a variety of concerns and problems. Three species—bighead, silver, and black carp—are of particular note, based on the perceived degree of environmental concern. Current controversy relates to what measures might be necessary and sufficient to prevent movement of Asian carp from the Mississippi River drainage into the Great Lakes through the Chicago Area Waterway System. Bills have been introduced in the 111th Congress to direct actions to avoid the possibility of carp becoming established in the Great Lakes.

Chapter 2- Two species of carp – the bighead carp and silver carp – hereinafter referred to as Asian carp, were first introduced into the southern United States in the early 1970s to remove algae from farmed catfish ponds. Flooding in the early 1990s caused many of the catfish farm ponds to overflow, and Asian carp were released into local waterways in the Mississippi

River basin. The carp have since migrated northward up the Mississippi River, becoming the most abundant species in some areas of the River.[1]

Chapter 3- The administration continues to make restoration and protection of the Great Lakes a national priority as evidenced by President Obama's, significant investment in the ecosystem under his Great Lakes Restoration Initiative (Initiative).

EPA understands the extreme level of concern that the public feels for the ecosystem, their safety while recreating, and for their jobs. We also have an urgent need to keep Asian carp from becoming established in the Great Lakes. As we move forward, we are working to keep Asian carp from becoming established in Lake Michigan. But to do that will require a coordinated, cooperative approach.

Chapter 4- Thank you for the opportunity to testify about ongoing efforts to address the risk to the Great Lakes posed by the migration of two species of Asian carp, the silver and bighead, through the Chicago Area Waterway System. Asian carp represent a grave threat to the Great Lakes fisheries and aquatic resources, including those managed by the National Park Service in areas such as Sleeping Bear Dunes National Lakeshore, Isle Royale National Park and Apostle Islands National Lakeshore. The Army Corps of Engineers (Corps) remains committed to use all available authorities, capabilities, and resources to combat this invasive species. The Corps cannot do this on its own, but continues to work intensively with, and leverage the full capabilities of Federal, State, provincial, bi-national, and municipal agency partners. I would like to briefly describe the Corps of Engineers' role in this important effort, current actions that the Corps is taking, plans for the immediate future and near term, as well as the Corps' longer term strategy.

Chapter 5- Our commitment to this task has been and remains unwavering. We have been working closely contributed $1.8 million to this effort. Most recently Illinois DNR served as the lead agency for the successful Rapid Response effort last December to prevent the migration of Asian carp when the electric barrier system was shut down for maintenance. The unified response of the Great Lakes States and Provinces was a shining leadership moment for our region, and a prime example of how a small group of committed people can make a difference.

Chapter 6- I appreciate the opportunity to testify today about the looming catastrophe that we face if Asian carp become established in the Great Lakes. I also appreciate the members of the Michigan Congressional delegation joining us today and for their past work on this and other Great Lakes issues.

I have been a conservation professional for over 30 years, and my role within the Michigan Department of Natural Resources and Environment is to protect our resources while maximizing recreational opportunities. Allowing Asian carp to populate our Great Lakes will destroy the resource as well as recreational opportunities. We must act swiftly, collaboratively, and wisely to address this crisis.

Chapter 7- The Great Lakes are a treasure of international significance. They contain approximately 20% of the world's surface freshwater, and 95% of North America's. One in three Canadians and one in 10 U.S. residents depend on the Great Lakes for their water. More than *35 million U.S. residents* and 8 million Canadians live, work, and recreate in, on or by the waters of the Great Lakes Basin.

Chapter 8- I am a biologist. I received a D.Phil. from the University of Oxford as a Rhodes Scholar. I come to the issue of Asian carp in the Great Lakes from the perspective of an active researcher on many kinds of invasive species in many places worldwide. I have been working on invasive species for 26 years, supported by private funding as well as funding from the National Science Foundation, the Environmental Protection Agency, the National Oceanograpic and Atmospheric Administration, and, of greatest relevance to today's hearing, the Army Corps of Engineers. I have co-edited two books and have authored or co-authored at least 150 published scientific papers. At the University of Notre Dame, I am the Director of the Center for Aquatic Conservation, Director of Notre Dame's new Environmental Change Initiative, and a professor of biology. My colleagues, collaborators, and I have many on-going research projects on various aspects of invasive species. The topics of particular relevance to this hearing include: (a) forecasting the spread and the environmental and economic impact of many aquatic nuisance species; (b) developing new species risk assessment (screening) protocols to allow private sector and agencies to prevent the importation and sale of harmful species while leaving unaffected the sale of benign species; (c) combining economic and ecological risk analyses to guide allocation of resources among management options; and (d) developing new genetically-based surveillance technologies for environmental protection.

Chapter 9- The Great Lakes are an extremely valuable resource for both the United States and Canada. The Great Lakes' commercial, recreational, and tribal fisheries are valued at more than $7 billion annually (ASA 2008). The lakes provide drinking water for 40 million people and are a rich tourist draw. They are a way of life for the people of the region and a healthy, vibrant Great Lakes ecosystem is immeasurable in economic terms alone.

Chapter 10- Good afternoon Madame Chairman. I am Del Wilkins, Vice President of Terminal Operations & Business Development for Canal Barge Company. Canal Barge Company is a family-owned business headquartered in New Orleans, Louisiana that has been in business for 76 years, and operates throughout the inland waterways system, from the Texas coast up to Pittsburgh and Chicago. Canal Barge owns Illinois Marine Towing, a Chicago-area towing and barge fleeting company, and Canal Terminal Company, a bulk liquid storage facility just south of Joliet, Illinois. Canal Barge and its affiliates are daily users of the Illinois waterways and bring both liquid and dry cargo in and out of the Chicago area from all over the country. I am pleased to testify today on behalf of the American Waterways Operators, the national trade association for the tugboat, towboat and barge industry. Thank you for the opportunity to be here and share our concerns about the manner in which government addresses the serious issue of the spread of Asian carp into the Great lakes.

Chapter 11- Good afternoon Madame Chairman. I am Del Wilkins, Vice President of Terminal Operations & Business Development for Canal Barge Company. Canal Barge Company is a family-owned business headquartered in New Orleans, Louisiana that has been in business for 76 years, and operates throughout the inland waterways system, from the Texas coast up to Pittsburgh and Chicago. Canal Barge owns Illinois Marine Towing, a Chicago-area towing and barge fleeting company, and Canal Terminal Company, a bulk liquid storage facility just south of Joliet, Illinois. Canal Barge and its affiliates are daily users of the Illinois waterways and bring both liquid and dry cargo in and out of the Chicago area from all over the country. I am pleased to testify today on behalf of the American Waterways Operators, the national trade association for the tugboat, towboat and barge industry. Thank you for the opportunity to be here and share our concerns about the manner in which government addresses the serious issue of the spread of Asian carp into the Great lakes.

Chapter 12- The dispersal barriers are located in the Chicago Sanitary and Ship Canal (CSSC), which is a man-made waterway creating the only continuous connection between Lake Michigan and the Mississippi River basin. The dispersal barrier system was developed to prevent the spread of invasive fish species between these watersheds.

In: Troubled Waters: Asian Carp and the... ISBN: 978-1-61761-766-9
Editors: Deborah M. Maeda

Chapter 1

ASIAN CARP AND THE GREAT LAKES REGION

Eugene H. Buck, Harold F. Upton, Charles V. Stern and James E. Nichols

SUMMARY

Four species of non-indigenous Asian carp are expanding their range in U.S. waterways, resulting in a variety of concerns and problems. Three species—bighead, silver, and black carp—are of particular note, based on the perceived degree of environmental concern. Current controversy relates to what measures might be necessary and sufficient to prevent movement of Asian carp from the Mississippi River drainage into the Great Lakes through the Chicago Area Waterway System. Bills have been introduced in the 111th Congress to direct actions to avoid the possibility of carp becoming established in the Great Lakes.

According to the Great Lakes Fishery Commission, Asian carp pose a significant threat to commercial and recreational fisheries of the Great Lakes. Asian carp populations could expand rapidly and change the composition of Great Lakes ecosystems. Native species could be harmed because Asian carp are likely to compete with them for food and modify their habitat. It has been widely reported that Great Lakes fisheries generate economic activity of approximately $7 billion annually. Although Asian carp introduction is likely to modify Great Lakes ecosystems and cause harm to fisheries, studies forecasting the extent of potential harm are not available. Therefore, it is not possible to provide estimates of potential changes in the regional economy or economic value (social welfare) by lake, species, or fishery.

The locks and waterways of the Chicago Area Waterway System (CAWS) have been a focal point for those debating how to prevent Asian carp encroachment on the Great Lakes. The CAWS is the only navigable link between the Great Lakes and the Mississippi River, and many note the potential of these waterways to facilitate invasive species transfers from one basin to the other. The U.S. Army Corps of Engineers has constructed and is currently operating electrical barriers to prevent fish passage. However, in light of recent indications that Asian carp may be present upstream of the barriers and in Lake Michigan, increased federal funding to prevent fish encroachment has been announced by the Obama Administration, and calls to permanently separate the two basins have grown. The potential closure of existing navigation structures in the CAWS and the permanent separation of the basins are currently the most contentious issues related to Asian carp control in the region, and a long-term solution has yet to be decided.

On January 19, 2010, the Supreme Court refused to order emergency measures sought by the State of Michigan to stop the migration of invasive Asian carp toward Lake Michigan from rivers and a sanitary canal in Illinois. Without comment, the Court refused to issue a preliminary injunction that would have closed waterway locks and required other temporary measures in reaction to the discovery of Asian carp upstream in Illinois rivers. On February 4, 2010, Michigan's Attorney General Mike Cox filed a renewed motion, asking the Supreme Court to reconsider issuing a preliminary injunction for the closure of Chicago-area locks based on new evidence that Asian carp are present in Lake Michigan. Michigan's renewed motion for a preliminary injunction was denied by the Supreme Court on March 22, 2010.

In the 111th Congress, Section 126 in Title I of P.L. 111-85 directed the U.S. Army Corps of Engineers to implement additional measures to prevent invasive species from bypassing the Chicago Sanitary and Ship Canal Dispersal Barrier Project and dispersing into the Great Lakes. Other bills have been introduced to list additional Asian carp species as injurious under the Lacey Act (H.R. 48, H.R. 3173, S. 237, S. 1421), and to direct various federal agencies to take specific actions to increase control over and restrict the spread of Asian carp (H.R. 51, H.R. 4472, S. 237, S. 2946).

BACKGROUND

Four species of non-indigenous Asian carp are expanding their range in U.S. waterways, resulting in a variety of concerns and problems. Three species—bighead, silver, and black carp—are of particular note, based on the perceived degree of environmental concern. Current controversy relates to what measures might be necessary and sufficient to prevent movement of Asian carp from the Mississippi River drainage into the Great Lakes through the Chicago Area Waterway System. Movement of Asian carp into the Great Lakes is ultimately of concern because increased numbers of carp in the Great Lakes increases the risk that Asian carp will establish reproducing populations in these waters. Bills have been introduced in the 111th Congress to direct actions to avoid the possibility of carp becoming established in the Great Lakes.

Grass Carp[1]

The grass carp or white amur, *Ctenopharyngodon idella*, was first imported to the United States in 1963 by the U.S. Fish and Wildlife Service for biological control of vegetation in aquatic environments. Grass carp are stocked to biologically control invasive aquatic plants, such as *Hydrilla* and Eurasian water milfoil. Shallow, quiet waters are their typical habitat, and this species easily tolerates waters near freezing. Grass carp initially escaped from the U.S. Fish and Wildlife Service Fish Farming Experimental Station in Stuttgart, AR. By 1970, grass carp had been stocked in lakes and reservoirs throughout the southeast United States and in Arizona, including some that were open to stream systems.[2] It has since spread widely across the country (**Figure 1**), including to four of the Great Lakes. Most grass carp now are stocked as sterile triploids,[3] and grass carp have not established breeding populations in the Great Lakes basin.

Black Carp[4]

The black carp, *Mylopharyngodon piceus*, arrived in the United States in 1973 with silver and bighead carp. Subsequently, this species was imported as a food fish, as the only cost-effective biological control agent to control non-native snails in catfish aquaculture ponds in Arkansas and Mississippi, and as a potential sterile biological control agent for zebra mussels. Of the four species of carp in U.S. waterways, black carp has the most limited known distribution (**Figure 2**).

The preferred habitat of black carp is along the bottom in deep water of large rivers. Owing to this habitat preference for deeper waters, sampling to determine black carp distribution is considered incomplete, since sampling is more difficult in deeper waters. Black carp feed primarily on mussels and snails, and there are concerns that black carp may harm native mollusks, many of which are listed as threatened or endangered under the Endangered Species Act.

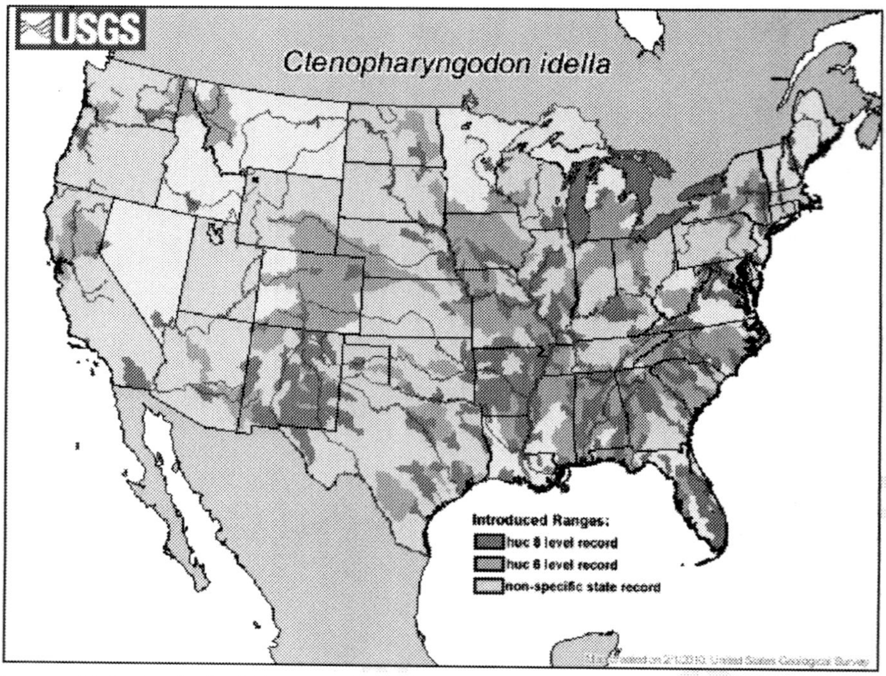

Source: U.S. Geological Survey, Nonindigenous Aquatic Species Fact Sheet on grass carp.

Notes: HUC (Hydrologic Unit Code) indicates to how much of a drainage basin the data apply. HUC 6 = one or more grass carp captured in the drainage basin. HUC 8 = one or more grass carp captured in the drainage subbasin. Records should not be interpreted as indicating the current presence of grass carp in all these areas.

Figure 1. Records of Grass Carp Capture, as of February 1, 2010

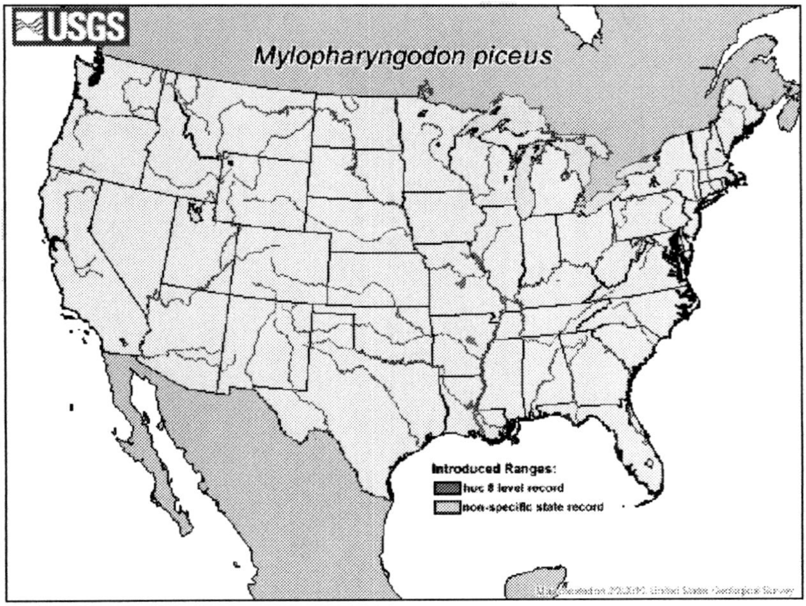

Source: U.S. Geological Survey, Nonindigenous Aquatic Species Fact Sheet on black carp.

Notes: HUC 8 = one or more black carp captured in the drainage subbasin. Records should not be interpreted as indicating the current presence of black carp in all these areas.

Figure 2. Records of Black Carp Capture, as of February 2, 2010

Silver Carp[5]

Silver carp, *Hypophthalmichthys molitrix*, were brought into the United States in 1973 under an agreement of maintenance between a private fish farmer and the Arkansas Game and Fish Commission.[6] This species has been used to control phytoplankton (microscopic drifting algae) in nutrient-rich water bodies and is also a food fish. Escapes from a state fish hatchery and from research projects involving use of these fish in municipal sewage systems,[7] as well as possible inclusion of silver carp among other fish shipments, contributed to the spread of this species. Silver carp proved unsuitable for U.S. aquaculture, and were never widely used. The U.S. distribution of silver carp is confined primarily to the Mississippi River drainage, with no record of capture in the Great Lakes (**Figure 3**).

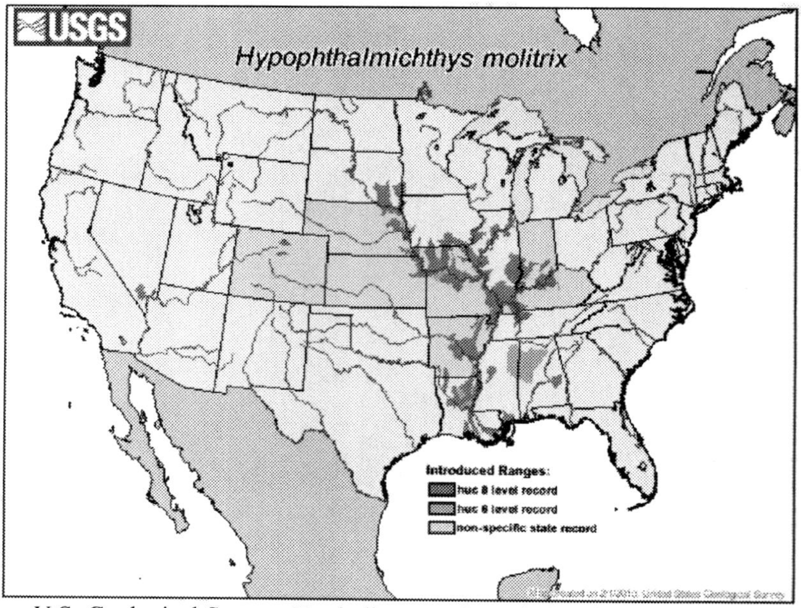

Source: U.S. Geological Survey, Nonindigenous Aquatic Species Fact Sheet on silver carp.

Notes: HUC (Hydrologic Unit Code) indicates to how much of a drainage basin the data apply. HUC 6 = one or more silver carp captured in the drainage basin. HUC 8 = one or more silver carp captured in the drainage subbasin. Records should not be interpreted as indicating the current presence of silver carp in all these areas.

Figure 3. Records of Silver Carp Capture, as of February 1, 2010

The silver carp is a filter-feeder, capable of consuming large amounts of phytoplankton, zooplankton (small drifting and/or swimming invertebrates), and detritus. Silver carp are easily startled by outboard motors, causing them to jump several feet out of the water.

Bighead Carp[8]

Bighead carp, *Hypophthalmichthys nobilis*, were brought into the United States in 1973 under an agreement of maintenance between the Arkansas Game and Fish Commission and a private fish farmer.[9] They proved suitable for U.S. aquaculture[10] and continue to be economically important in Arkansas, Mississippi, and Alabama.[11] This species was discovered in open waters of the Ohio and Mississippi Rivers in the 1980s, probably after escaping from fish hatcheries and/or research projects involving use of these fish in municipal sewage systems. [12] In the United States, bighead carp are found primarily in the Mississippi River drainage. However, a limited number of bighead carp were captured by commercial fishermen in Lake Erie between 1995 and 2003 (**Figure 4**).

Like silver carp, bighead carp typically require large rivers for spawning, but inhabit lakes, backwaters, reservoirs, and other low-current areas during most of their life cycle. They are filter- feeders, consuming primarily phytoplankton and zooplankton.

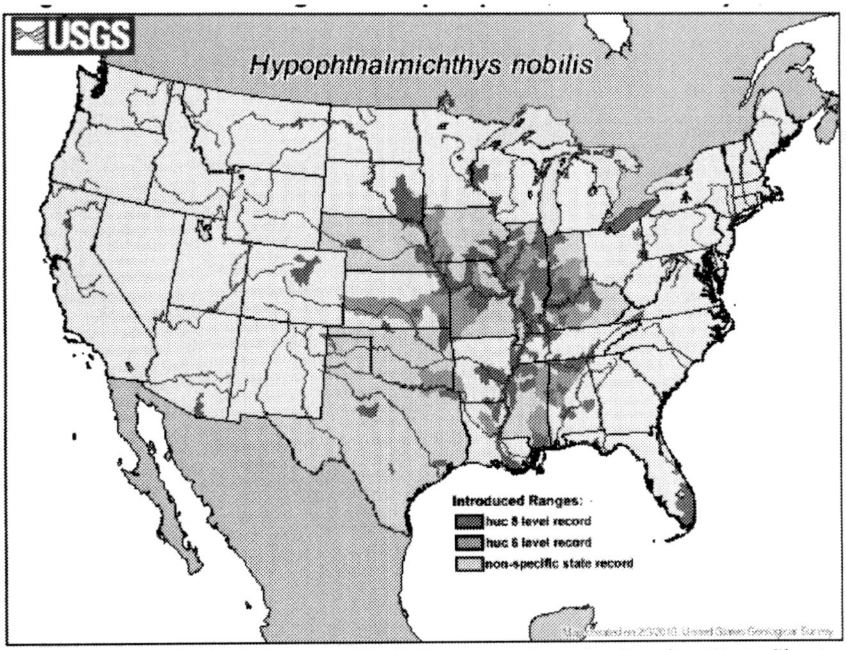

Source: U.S. Geological Survey, Nonindigenous Aquatic Species Fact Sheet on bighead carp.

Notes: HUC (Hydrologic Unit Code) indicates to how much of a drainage basin the data apply. HUC 6 = one or more bighead carp captured in the drainage basin. HUC 8 = one or more bighead carp captured in the drainage subbasin. Records should not be interpreted as indicating the current presence of bighead carp in all these areas.

Figure 4. Records of Bighead Carp Capture, as of February 3, 2010

Managing Non-Native Species

Non-native species that do become established commonly exist at low populations for several generations, after which some begin a period of rapid population growth and range expansion. Although initial captures of wild silver carp were reported in the early 1970s, silver carp only rarely were captured in U.S. rivers until about 1999, after which their population began to grow at an exponential rate. Some suggest that floods in the early 1990s may have provided excellent spawning and recruitment opportunities for silver carp, and stimulated their later exponential growth phase.[13] Field experience in the United States has shown that silver carp generally follow a few years after bighead carp in colonizing new habitat.[14]

Many factors may contribute to the introduction and spread of non-native species. For example, juvenile silver and bighead carp are easily mistaken for native baitfish. Thus, the dumping of unused bait by sport fishermen may contribute to the introduction and spread of these species. In addition, bighead carp (as well as a number of other potentially invasive non-native fish species) have been reared, transported, and traded in large numbers as live fish for human food, especially in large metropolitan areas. Such commerce in bighead carp occurred with relatively limited state and local regulation until recently.

Eradication of non-native species in aquatic environments is difficult and rare, having only occasionally been successful when efforts were focused on small-scale and closed systems like reservoirs, ponds, small locks, and marinas. Since eradication of a non-native species, once it has become established, is unlikely, difficult, and therefore expensive, management more often focuses on preventing troublesome species for entering new habitats, through regulating imports of certain nuisance species, preventing or slowing the spread of already introduced species, and monitoring to detect new invaders when their populations may be localized and at low densities such that eradication might still be possible. [15] While efforts to prevent introduction may be costly, it almost always will be less expensive than continued attempts to eradicate or control non-native species that become established.

POTENTIAL IMPACTS

Ecological Concerns

According to the Great Lakes Fishery Commission,[16] Asian carp pose a significant threat to fisheries of the Great Lakes.[17] Asian carp populations could expand rapidly and change the composition of Great Lakes ecosystems. Direct ecological effects are likely to result from their various diets: silver carp eat phytoplankton, bighead carp eat zooplankton, black carp eat invertebrates such as snails and mussels, and grass carp eat aquatic plants. Resident Great Lakes fish species could be harmed, because Asian carp are likely to compete with them for food and modify their habitat. Species at greatest risk include native mussels, other aquatic invertebrates, and fishes.[18] As bighead and silver carp have dispersed and migrated within the Mississippi River drainage, these species have out-competed native fish to become the most abundant fish in certain areas.[19]

On the other hand, others have predicted that black carp are not likely to become established in the Great Lakes if introduced, while silver carp are predicted neither to spread quickly nor to be perceived as a nuisance in the Great Lakes.[20] Bighead carp were not considered in this analysis.

Furthermore, the Great Lakes today are hardly pristine habitat, with the intentional human introduction of non-native species (e.g., brown and rainbow trout, coho and Chinook salmon) characterizing fishery management of the waters for many years. The intentional and accidental introduction of non-native species has changed this historic ecosystem in many ways, including depletion of previously dominant lake trout and whitefish species. In addition, the ecological changes wrought by non-native species arriving in ship ballast water (e.g., zebra mussels, round goby) and by other means (e.g., lamprey and alewife) have been substantial.

Economic Concerns

Recreational and commercial fisheries of the Great Lakes depend on fish populations that could be affected by Asian carp. The primary economic impacts of Asian carp are likely to be related to these fisheries, although concerns have also been raised about potential effects on recreational boating and hunting.[21] Although the net effects are likely to be negative, it is also possible that the introduction of Asian carp to the Great Lakes may provide some utility such as the development of new commercial and recreational fisheries.[22]

It has been widely reported that Great Lakes fisheries generate U.S. economic activity of approximately $7 billion annually.[23] One should exercise caution in using this figure for assessing public policy alternatives or to make comparisons with the value of other economic sectors. The Great Lakes is composed of many fisheries, each specific to different water bodies, species, and groups of users. Asian carp are likely to affect each lake and areas within lakes to varying degrees because of different biological, chemical, and physical conditions. Anglers will be affected to different degrees depending on local ecological interactions and substitute angling opportunities.

Measures of economic activity such as the $7 billion of economic impacts are only one dimension of economic analysis. The economic input-output studies of the recreational and boating sectors provided below cannot be used to estimate changes in social welfare,[24] to assess trade-offs among public policy alternatives, or to conduct benefit-cost analysis. To more fully understand how society would be affected, valuation studies would be required to estimate the potential changes in social welfare resulting from Asian carp introduction.

Although Asian carp introduction is likely to harm many Great Lakes fisheries, potential changes to ecosystems and the associated economy are not well understood. It is questionable whether accurate predictions of changes by lake, species, and associated fishery are possible. Potential changes resulting from species invasions are difficult to assess because of the underlying complexity of ecological and economic systems. Data and models required to make these assessments are not available and complete assessments would be costly and likely require years of research. The lack of definitive predictions does not mean that the effects of Asian carp introduction would not be significant or that managers should wait to assess the actual effects as Asian carp become established in the Great Lakes. Existing information related to Asian carp movement and population increases in the Mississippi Basin and the magnitude of recreational activities in the Great Lakes indicate that a major threat exists and the effects are likely to be significant.

The economic contributions of recreational and commercial activities on state and regional economies of the Great Lakes region are significant. The economic input-output data cited below measure financial activities associated with the money people spend to buy goods and services on their fishing trips. Expenditures at businesses that provide goods and services have direct, indirect, and induced effects on business revenues, jobs, and personal income in the local area and at the state level. This approach to assessing recreational fishing is the expenditure and economic impact approach. The following descriptions provide recent economic information, but do not consider the effects of Asian carp introduction.

The Great Lakes' recreational fisheries target perch, black bass, walleye, lake trout, salmon, pike, steelhead, and others. In 2006, approximately 1.5 million anglers fished 17.9 million recreational days on the Great Lakes.[25] These anglers spent an estimated $1.2 billion during Great Lakes fishing trips and $1.3 billion on equipment for activities related to Great Lakes fishing.[26]

Table 1. Great Lakes Recreational Fishing Activity and Economic Impacts in 2006

States	Anglers	Days Fished	Retail Sales (000s)	Salaries (000s)	Jobs	Total Impact (000s)
Illinois	56,000	728,000	$93,589	$55,158	1,511	$175,074
Indiana	46,000	759,000	$224,588	$117,321	4,170	$394,866
Michigan	461,000	6,981,000	$562,654	$312,197	8,283	$1,001,641
Minnesota	48,000	272,000	NR	NR	NR	NR
New York	247,000	2,060,000	$213,174	$122,147	3,288	$369,194
Ohio	328,000	2,807,000	$480,482	$248,301	9,915	$801,817
Pennsylvania	85,000	598,000	$399,342	$213,921	5,200	$725,705
Wisconsin	235,000	3,705,000	$315,336	$159,420	6,153	$528,274
Totals (Great Lakes States)	**1,506,000**	**17,910,000**	**$2,289,165**	**$1,228,465**	**38,520**	**$3,996,571**
Totals (United States)			**$2,524,266**	**$2,189,490**	**58,291**	**$7,089,230**

Sources: U.S. Department of the Interior, Fish and Wildlife Service, and U.S. Department of Commerce, Census Bureau, *2006 National Survey of Fishing, Hunting, and Wildlife-Associated Recreation*, Washington, DC, 2007. Southwick Associates, *Sportfishing in America: An Economic Engine and Conservation Powerhouse*, American Sportfishing Association, Multistate Conservation Grant Program, 2007.

Notes: Great Lakes fishing includes lakes Superior, Michigan, Huron, Ontario, Erie, and St. Clair, connecting waters, and fishing in tributaries for smelt, steelhead, and salmon. Minnesota economic impacts were not reported (NR) because of small sample size. Illinois (<10), Indiana, and Pennsylvania estimates should also be used with caution because of small sample sizes (10 to 30). Retail sales include trip and equipment expenditures. Equipment expenditures were prorated according to how and where equipment such as boats were used. United States totals include economic impacts outside Great Lakes states that resulted from trip and equipment expenditures for Great Lakes fishing.

Economic impacts resulting from these expenditures included more than 58,000 jobs, salaries of $2.1 billion, and total impacts[27] throughout the U.S. economy of slightly more than $7 billion.[28] Great Lakes fisheries also support charter boat fishing businesses that provide recreational fishing services to anglers. In 2002, an estimated 1,746 charter firms made more than 93,000 charter trips in the Great Lakes region.[29] **Table 1** provides a breakdown of angling activity and economic impacts of recreational fishing by state.

In 2008, Great Lakes commercial fishing produced 18.3 million pounds of fish with a landed value[30] of nearly $17 million (**Table 2**).[31] Commercial fisheries are important to many coastal communities, and except for Lake Erie, each lake supports tribal fisheries. Top species are lake whitefish, yellow perch, walleye, chubs, and smelt. For certain species, specific lakes contribute the bulk of commercial landings—including Lake Huron (60% of whitefish), Lake Erie (84% of yellow perch, and 94% of smelt), and Lake Michigan (80% of chubs).[32] Record harvests occurred in 1899, when 120 million pounds were landed in the United States.[33] Landings were dominated by lake herring and chubs (64 million pounds), lake trout (10 million pounds), and yellow perch (10 million pounds).[34] Landings and value of commercial fisheries in the Great Lakes have declined dramatically because of factors such as invasive species, pollution, habitat degradation, overfishing, competition with imports, personal tastes and preferences, and regulatory changes.

There are 4.3 million boats registered in the Great Lakes states, and it has been estimated that 911,000 operate on the Great Lakes.[35] When disturbed by a boat motor, silver carp may jump as high as 10 feet out of the water. In parts of the Mississippi River drainage, silver carp have caused injuries and damaged equipment when large fish have jumped into moving boats. Silver carp also could injure boaters and water-skiers and detract from boating in the Great Lakes. As in the case of fisheries, predictions of the potential magnitude of economic effects on Great Lakes boating are not available.

In 2004, the U.S. Army Corps of Engineers in partnership with the Great Lakes Commission undertook a study of recreational boating in the Great Lakes states. Recreational boaters spent approximately $9.8 billion during trips and $5.7 billion on craft in Great Lakes states.[36] Economic results from these expenditures included more than 246,000 jobs and salaries of $6.5 billion. **Table 3** provides economic measures of boating on Great Lakes states. The study found that a significant share of boating expenditures took place at Great Lakes marinas. It is also likely that a significant portion of boating expenditures are related to fishing activity.

Table 2. Great Lakes Commercial Fishing Landings and Revenue in 2008

State	Landings (pounds)	Revenue
Michigan	9,998,000	$7,448,000
Minnesota	318,000	$158,000
New York	44,000	$65,000
Ohio	4,493,000	$5,315,000
Pennsylvania	50,000	$140,000
Wisconsin	3,376,000	$3,641,000
Total	**18,279,000**	**$16,767,000**

Source: U.S. Department of Commerce, National Marine Fisheries Service, *Fisheries of the United States 2008*, Silver Spring, MD, July 2009. p.6.

Table 3. Annual Economic Impact of Boating on Great Lakes States in 2003 (Includes All Registered Boats and Boating in Great Lakes States)

State	Boats (000s)	Sales (000s)	Jobs	Salaries (000s)
Illinois	360,252	$1,958,000	22,407	$678,000
Indiana	216,145	$2,203,000	30,437	$710,000
Michigan	953,554	$3,905,000	51,329	$1,342,000
Minnesota	845,094	$3,709,000	49,060	$1,247,000
New York	528,094	$2,749,000	28,901	$987,000
Ohio	413,048	$1,959,000	26,148	$656,000
Pennsylvania	355,235	$71,000	1,195	$24,000
Wisconsin	610,800	$2,493,000	36,640	$825,000
Total	**4,282,222**	**$19,047,000**	**246,117**	**$6,479,000**

Source: Great Lakes Commission, *Great Lakes Recreational Boating's Economic Punch*, Ann Arbor, MI, 2004.

Social Concerns

The introduction of Asian carp to the Great Lakes, potentially changing lake ecosystems from "salmon and trout dominated" to "carp dominated," has the potential to damage the public image of these lakes and to lower the feeling of "well being" and pride of area residents.[37] As such, the introduction of these species could reduce the social value of lake-related activities.

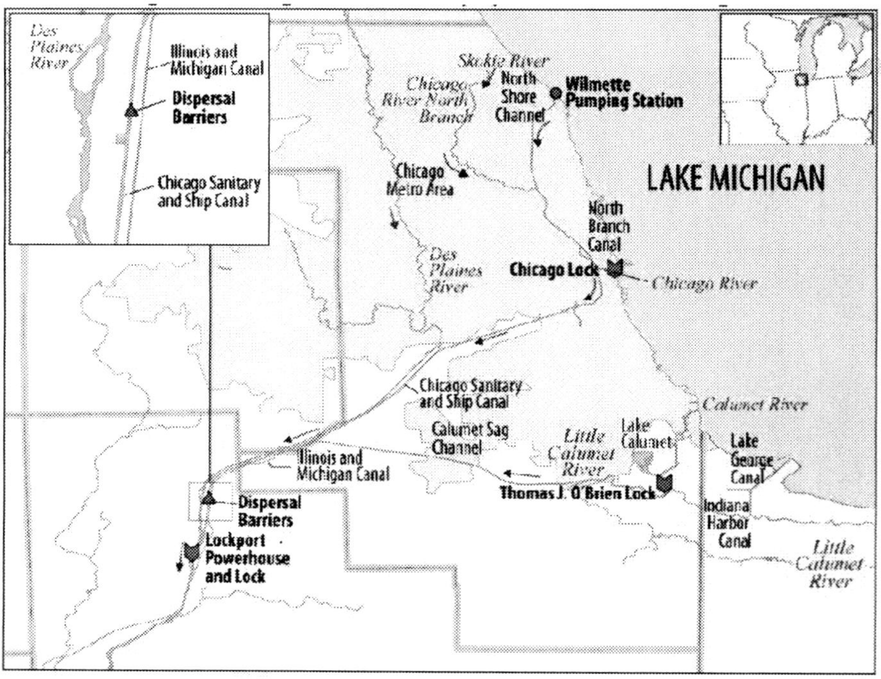

Source:. Adapted by the Congressional Research Service, February 2010.

Figure 5. Chicago Area Waterway System and Lake Michigan

THE CHICAGO AREA WATERWAY SYSTEM (CAWS)

The Chicago Area Waterway System (CAWS) is a segment of the Illinois Waterway in northeastern Illinois and northwestern Indiana. The Illinois Waterway is a 327-mile channel running from Chicago to St. Louis. It is maintained at a minimum depth of 9 feet by the U.S. Army Corps of Engineers (hereinafter referred to as the Corps).[38] It is the only navigable link between two of the largest freshwater drainage basins in the world, the Great Lakes and the Mississippi River. The CAWS portion of the Illinois Waterway includes modified rivers, locks, canals and other structures that control the flow of water through the Chicago metropolitan area. It has recently received attention for its potential to provide a pathway for Asian carp to migrate from the Mississippi River and its tributaries into the Great Lakes. The system of projects comprising the CAWS is shown in **Figure 5**.

Historically, an important geologic feature in the Chicago area's watershed was the Chicago Portage. The Chicago Portage separated the drainage basins of the Mississippi River and the Great Lakes prior to modification of these waterways. These bodies of water were first artificially connected for navigation in 1848 through a privately-constructed 97-mile canal connecting the Chicago River to the Illinois River. This canal, known as the Illinois and Michigan (I&M) Canal, was maintained for commercial use from 1848 to 1933. It was eventually replaced by the network of canals and locks that comprises the CAWS.[39] Canals within the CAWS today include the Chicago Sanitary and Ship Canal (or CSSC, completed in 1900), the North Shore Channel (completed in 1910) and the Cal-Sag Channel (completed in 1922). During construction of these canals, the flows of the Chicago River and the Calumet River were also permanently reversed away from Lake Michigan and toward the Mississippi River drainage basin through structural modifications and pumping.[40] The altered flow of the rivers prevented sewage discharge into the canals from contaminating Chicago's drinking water supply intakes on Lake Michigan.

In recent years, the locks of the CAWS have become a focal point for those debating how to prevent invasive species (and specifically, Asian carp) encroachment between the Great Lakes and the Mississippi River. The Corps operates multiple lock sites that connect the CAWS to the Great Lakes, including the O'Brien Lock (on the Cal-Sag Channel) and the Chicago Lock (on the Chicago River; see **Figure 5**). Both of these locks include sluice gates operated by the Metropolitan Water Reclamation District of Greater Chicago (MWRD) that can provide flood control in severe rainstorms.[41] The MWRD independently owns and operates a third site (the Wilmette pumping station) on the North Shore Channel that directly connects the CAWS to the Great Lakes. The Corps also owns and operates the lock at Lockport Powerhouse and Lock, which is southwest of Chicago on the CSSC. (See **Figure 5**.) Due to its distance from the Great Lakes and the fact that the Corp's electric fish barriers operate upstream on the CSSC, this third lock has not been as prominent in recent invasive species debates.

The CAWS plays a significant role in the region's commercial and recreational navigation, although estimates of the full economic value of the locks within the CAWS (in particular, O'Brien Lock) vary widely. The Chicago Lock, one of the country's busiest locks for traffic, handled 36,256 vessels and conducted 11,599 lockages in 2008.[42] The O'Brien Lock handled 17,532 vessels and conducted 6,310 lockages in 2008.[43] While most of the traffic on the Chicago Lock is recreational, the transit of commodity-laden commercial barges is higher at O'Brien Lock, which allows for shippers to offload onto deepwater vessels.[44] Statistics from the Corps indicate that approximately 7 million tons worth of commodities move through O'Brien lock annually, including bulk quantities of sand and gravel, coal, and steel.[45]

Additional analysis, including a comparison of alternative means of freight transit, is necessary to fully understand the value of the locks to the region. In response to an initial estimate by the Corps that shippers saved approximately $192 million by using the O'Brien and Chicago locks in 2008 (or an addition of approximately $27 per ton of freight shipped), the State of Michigan commissioned a study which concluded that lock closure would lead to considerably lesser impacts than estimated by the Corps. The Michigan study estimated that a shift from barge to overland shipping would result in additional costs of approximately $64 million-$69 million annually, or approximately $10 per ton.[46] This study has since been criticized by the Illinois Chamber of Commerce, which has published several academic critiques of the Michigan study, as well as a separate study estimating a much higher cost associated with lock closure.[47] In contrast to the Michigan Study, the Illinois Chamber of Commerce study estimated a total cost of $530 million-$580 million annually over the next eight years for lock closure, and a net cost to the Chicago economy of $4.7 billion over a 20-year horizon.[48] The two studies differ considerably in their treatment of several important assumptions, including those related to both direct and indirect costs for the transition to overland shipping in the areas around the locks.

FEDERAL RESPONSE TO ASIAN CARP

Federal responses to the introduction of Asian carp in U.S. waters can generally be divided into two categories. First, Congress has directed the Corps and other agencies to undertake a limited number actions that relate to the need to block inter-basin passage of aquatic nuisance species in the CAWS (both upstream and downstream). Since 2001, Asian carp prevention has been the most high-profile of these efforts, and has been conducted by the Corps with planning coordination and funding from other agencies. Second, the federal government is engaged in long-term, nationwide planning and management of Asian carp under authorities codified in the Nonindigenous Aquatic Nuisance Prevention and Control Act of 1990 (P.L. 101-646, as amended) and other statutes.[49] These actions have usually been conducted by the Aquatic Nuisance Species Task Force (ANS Task Force), chaired by the Fish and Wildlife Service (FWS) and the National Oceanic and Atmospheric Administration (NOAA), with support provided by various other agencies, including the U.S. Environmental Protection Agency (EPA), the U.S. Geological Survey (USGS), and the Corps.

Short-Term Prevention Efforts in the CAWS

Electric Dispersal Barriers

In the National Invasive Species Act of 1996 (P.L. 104-332), Congress directed the Corps and the ANS Task Force to investigate environmentally sound methods to prevent the dispersal of aquatic nuisance species from the Great Lakes into the Mississippi River drainage.[50] In response, an advisory panel of federal, state, local, and international representatives (known as the Dispersal Barrier Panel) recommended an electronic dispersal barrier demonstration project at the southwestern end of the CSSC north of Lockport Powerhouse and Lock (see **Figure 5**) as the preferred short-term method to stop the movement of invasive species through the CAWS.[51] This type of barrier uses steel cables secured to the bottom of the canal to create a pulsating field of electricity that discourages fish from passing. It was selected based on projected cost, likelihood of success, environmental impacts, commercial availability, permit requirements, and effect on existing canal uses. The barrier was completed in 2001 and became operational in 2002.[52]

Around the same time the dispersal barrier became operational, rapid upstream encroachment of Asian carp toward Lake Michigan was becoming a management concern for the Fish and Wildlife Service. As a result, the demonstration barrier became the default method to prevent short-term encroachment for Asian carp. Based on subsequent experience and testing, the Dispersal Barrier Panel determined that the demonstration barrier should be upgraded into a stronger, more permanent barrier (Barrier I), and that construction of a second large barrier (Barrier II) would provide additional protection through redundancy in the barrier system.[53] Preliminary repairs to Barrier I were completed in October 2008, and the Corps plans to make Barrier I permanent and enhance its operating parameters after Barrier II is complete. Barrier II is located approximately 800 feet downstream from Barrier I, and has two sets of electrical arrays (known as Barriers IIA and IIB). Construction of Barrier IIA began in 2004, and this part of the barrier became permanently operational in April 2009 at a total cost of approximately $10 million. Barrier IIB is scheduled to be operating no later than October 2010, at an expected cost of approximately $13 million.[54] To date, the Corps has not estimated the long-term cost for barrier operations. However, the Corps requests $7.25 million for barrier operations in the President's FY201 1 Budget.[55]

Federal agencies have coordinated rapid response activities to supplement the barrier protection system through the Asian Carp Regional Coordinating Committee. This committee, led by EPA's Great Lakes Program, includes representatives from federal agencies, as well as some state, local, nongovernmental, and Canadian government organizations. To date, the most visible action by the committee have been chemical treatments on the CSSC (December 1-7, 2009) and the Little Calumet River (May 20-27, 2010). For the CSSC action, more than 450 individuals were involved in the mass rotenone treatment of a 5.7-mile stretch of the CSSC while Barrier IIA was taken down for scheduled maintenance. This effort located a single Bighead carp, 500 feet above the Lockport Powerhouse and Lock and downstream from the electric barriers.[56]

Dispersal Barrier Bypass Research

In the Water Resources Development Act of 2007 (WRDA 2007, P.L. 110-114), Congress directed the Corps to study other means to prevent the spread of Asian carp through the CAWS, including the range of options for technologies to prevent passage beyond the electrical dispersal barriers.[57] In response to this directive, the Corps initiated a number of studies. First, in January 2010, the Corps produced a study (known as the Interim I study) that recommended a network of concrete and chain link barricades to deter fish passage over the Des Plaines River during flooding or through culverts connecting the CSSC to the I&M canal.[58] This project is expected to be built with approximately $13.2 million in funding from the EPA's Great Lakes Restoration Initiative (GLRI), and is scheduled for completion by late October 2010. The Corps is also conducting a separate study (Interim II study) on optimal operating parameters for the electrical barriers. This final Interim II study is expected in late September 2010.

The Corps conducted a third study (Interim III study) exploring how its existing locks and other structures could be operated to minimize the likelihood of Asian carp infestation, and has convened meetings with navigation interests on potential operational changes for these structures. The Interim III study, released on June 2, 2010, concluded that partial changes in operating parameters would not be beneficial in slowing Asian carp migration;

however, the Corps plans to install fish screens on certain sluice gates and modify operations to provide lock closure during chemical and other control efforts.[59] An additional study (Interim IIIa study) focused on other deterrent measures that could be quickly employed to prevent Asian carp migration into the Great Lakes.[60] This study, completed in April 2010, concluded that a deterrent combining acoustic air bubble barrier technology and strobe lights (ABS deterrent) would be the best available measure to reduce Asian carp migration risk, and noted eight candidate sites at which the ABS deterrent could be utilized.

Monitoring

The Corps and other agencies, including the FWS, EPA, and USGS, are also contributing resources toward monitoring efforts to evaluate the presence and movements of Asian carp in the CAWS. In addition to conventional sampling methods such as electrofishing and netting, the Corps is working with the University of Notre Dame to conduct an experimental fish sampling method known as environmental DNA (eDNA) testing. This method filters water samples, then extracts fragments of shed DNA to search for genetic markers unique to Asian carp. The method has yet to complete independent peer review, but an audit of the methodology by EPA in February 2010 concluded that the technique is sufficiently reliable and robust in reporting a pattern of detection that should be considered actionable in a management context.[61] While few Asian carp have been located upstream of the barriers using conventional sampling methods, positive eDNA test results for Asian carp found in multiple locations upstream suggest that they are very likely present at multiple locations on the lake side of the barriers.

On June 23, 2010, the Asian Carp Regional Coordinating Committee announced the catch of a live Bighead carp at Lake Calumet (upstream of the electric barriers, between O'Brien Lock and Lake Michigan) by a fisherman under contract with the Illinois Department of Natural Resources.[62] The finding was significant as it represented the first live Asian carp located upstream of the barriers. It is currently unknown how the finding will affect previous management decisions, such as Corps decisions related to lock operations.

Nationwide Asian Carp Management and Long-Term Actions

Separate from efforts focusing on short-term prevention and other actions in the CAWS, the ANS task force has studied and initiated a number of nationwide management actions through its Asian Carp Working Group. Beginning around 2001, the working group requested and co-funded USGS risk assessments of multiple Asian carp species that found a high potential for black, silver, and bighead carp to become established in the United States.[63] In response to these findings, FWS listed black and silver carp as injurious under the Lacey Act in 2007.[64]

Also in 2007, FWS authored a study, *Management and Control Plan for Bighead, Black, Grass, and Silver Carps in the United States*, produced in collaboration with federal and non-federal stakeholders. The final plan outlines seven broad goals (divided into 133 short- and long-term recommendations) that would contribute to a goal of extermination of wild Asian carp. Recommendations in that report include a wide array of methods, including those intended to stop Asian carp encroachment (such as electric barriers, bubble curtains, and sonic barriers to control carp movement) as well as those that would eliminate wild Asian carp populations outright (including concentrated fishing operations, genetic manipulation, and pheromone baiting).[65]

Through its Midwest regional office, FWS is in the process of implementing some of these recommendations, although full implementation of the plan would require additional resources; FWS estimated that full implementation of all of the measures would cost at least $286 million over 20 years.[66] As outlined in the plan, other federal agencies, including the Corps, USGS, EPA, and the U.S. Coast Guard, would also contribute resources under their respective authorities.

The Corps is also studying the broader application of methods to permanently prevent interbasin transfer of aquatic nuisance species in this region. In WRDA 2007, Congress authorized a longterm study on the feasibility of approaches to permanently eliminate the risk of interbasin transfer of ANS, including permanent ecological separation of the basins.[67] This study, known as the Interbasin Transfer Feasibility Study, is expected to focus on the conveyance of Asian carp as well as other invasive species throughout the Great Lakes region (in addition to the CAWS). According to the Corps' FY2011 budget justifications for Civil Works, it has a remaining cost of $8.5 million. In 2008, the Alliance for the Great Lakes conducted its own preliminary study on ecological separation that highlighted major issues and pointed out specific research needs in this area.[68] Permanent ecological separation of the basins (especially in the Chicago area) would likely involve significant changes to existing navigation structures and operations in the CAWS, and is perhaps the most contentious issue related to the Asian carp debate. To date, no long-term solution has yet been decided. For more information on efforts to mandate permanent ecological separation, see the "Litigation" and "Congressional Interest" sections of this chapter.

Asian Carp Control Strategy Framework

Recent positive eDNA test results in the Great Lakes and the finding of a fish during the December 2009 rapid response chemical treatment of the CSSC have raised the profile of efforts to control Asian carp. On February 8, 2010, the White House convened a Summit for Great Lakes governors on the threat of Asian carp to the Great Lakes. This meeting focused on defining strategies to combat the spread of Asian carp and improving coordination and effective response across all levels of government. At this summit, the Obama Administration unveiled a draft framework that was subsequently finalized. This framework, known as the Asian Carp Control Strategy Framework, was last updated on May 5, 2010.[69] It outlines future actions and funding sources to eliminate the threat of Asian carp in the Great Lakes, and builds on the existing Corps barrier and monitoring as well as the 2007 FWS national management plan. The draft plan identifies 25 short- and long-term actions and $78.5 million in new funding ($58 million from the President's GLRI, funded by EPA) to implement these recommendations.[70] (See **Table 4**.)

Short-term recommendations in the framework were projected to be implemented by May 15, 2010. They include (1) ensuring proper supplies for future rapid response operations, including rotenone, netting, and personnel; (2) intensifying fish collection and other monitoring efforts (including eDNA); (3) modifying structural operations for locks, dams, sluice gates, and pumping stations; (4) expediting construction of the 13-mile barrier and sustained operations of Barrier IIA; and (5) researching applications of targeted biological controls (such as pheromone attractants). Long-term actions that are to be undertaken between now and 2020 include (1) feasibility studies of additional structural enhancements (electric and other barriers); (2) future rotenone applications; (3) implementation of biological controls; (4) sustained operations of barriers and continued exploration of monitoring techniques; and

(5) various other items, including controlled lock operations and development of a market for Asian carp.

Table 4. Asian Carp Control Strategy Matrix (Summary of Items with Funding Identified)

Action Item	Agency	Funding	Source	Start Date
Ongoing Barrier Operation	Corps	\$3,750,000	Corps	FY2010
Increase eDNA, Other Monitoring Capacity	Corps, IL DNR, FWS	\$3,540,000	GLRI, Corps, FWS, ILDNR	Ongoing
Barrier IIB Construction	Corps	\$13,000,000	Corps	2009
Implement Interim Study I/ Construct Other Barriers	Corps	\$13,200,000	GLRI	FY2010 Q-2
Final Report on Additional Barriers	Corps	\$1,100,000	Corps	2009
Inter-Basin Tran-sfer Feasibility Study, Lock Closure impacts	Corps	\$1,000,000	Corps, GLRI	FY2010
Commercial Market Enhancement	IL DNR	\$3,000,000	GLRI	FY2010
Additional Rotenone Actions	IL DNR, FWS, USCG	\$5,000,000	Not currently funded	FY2010 Q-4
Interbasin Transfer Assessment	USGS	\$500,000	GLRI	FY2010 Q-2
Targeted Removal	RR Team	\$2,000,000	GLRI	FY2010 Q-2
Commercial Fishing Removal	IL DNR, USCG	\$300,000	GLRI	FY2010
Lacey Act Enforcement	FWS	\$400,000	GLRI	FY2010

Action Item	Agency	Funding	Source	Start Date
Integrated Pest Management	FWS	$4,223,000	GLRI	FY2010
State Aquatic Nuisance Management Plans	FWS	$11,000,000	GLRI	FY2010
Activities to Support ANS Priorities	FWS	$8,500,000	GLRI & FWS	FY2010
Competitive Funding for Additional Support	EPA/FWS	$8,800.000	GLRI	FY2010
Research & Other Sciencea	USGS, multiple agencies	$4,203,000	Multiple Sources	Multiple start dates

Source: Asian Carp Workgroup, Asian Carp Control Strategy Matrix. (Adapted by CRS.)

Notes: This summary table combines multiple items in the Administration's framework. It also omits certain actions which do not require funding but were included in the framework.

a. Consists of 10 discrete items identified in the matrix to be undertaken by USGS and other agencies.

A review of the framework's recommendations indicates sustained or increased funding for most of the major ongoing federal efforts mentioned earlier in this chapter. Significantly, revisions to the framework since its original release have emphasized funding for the Interbasin Feasibility Studyand note 2012 as the targeted release date for the interim version of the study, which will focus on ecological separation of the CAWS. A more comprehensive study focusing on all Great Lakes waterways is to be completed subsequently.

LITIGATION

In an attempt to mitigate the movement of Asian carp into the Great Lakes, the state of Michigan filed suit in the Supreme Court against the state of Illinois, the Army Corps of Engineers, and the Metropolitan Water Reclamation District of Greater Chicago on December 21, 2009.[71] Michigan filed a motion for preliminary injunctive relief and a motion to reopen docket numbers 1, 2, and 3, Original, based on pre-existing litigation between Illinois and the Great Lakes states.[72] In its motion for preliminary injunctive relief, Michigan sought an order from the Court that would direct Illinois, the Army Corps of Engineers, and the Metropolitan Water Reclamation District of Greater Chicago to immediately close the shipping locks near Chicago and implement temporary emergency measures to prevent Asian carp from invading the Great Lakes.[73] Several other states bordering the Great Lakes supported Michigan's request for preliminary injunctive relief.[74] The Solicitor General, on behalf of the United States, filed a memorandum opposing Michigan's request for a preliminary injunction.[75]

Without comment, the Supreme Court issued an order on January 19, 2010, which summarily denied Michigan's request to close the shipping locks near Chicago.[76]

On February 4, 2010, Michigan's attorney general filed a renewed motion, asking the Supreme Court to reconsider issuing a preliminary injunction for the closure of Chicago-area locks based on new evidence that Asian carp are present in Lake Michigan.[77] Michigan's renewed motion for a preliminary injunction was denied—again without comment—by the Supreme Court on March 22, 2010.[78]

In addition to Michigan's request for preliminary injunctive relief, Michigan requested that the Supreme Court reopen docket numbers 1, 2, and 3, Original. In its motion to reopen, Michigan requested a supplemental decree from the Court declaring that the Chicago Waterway System constitutes a "public nuisance."[79] Additionally, Michigan requested that the Supreme Court grant a permanent injunction requiring Illinois, the Army Corps, and the Metropolitan Water Reclamation District of Greater Chicago "to expeditiously develop and implement plans to permanently and physically separate the carp-infested waters in the Illinois River basin" from Lake Michigan to prevent the spread of Asian carp into the lake. [80]

Michigan's motion to reopen docket numbers 1, 2, and 3, Original, however, was denied by the Supreme Court on April 26, 2010.[81] The Supreme Court also denied Michigan's request to file a new complaint with the Court.[82]

In response to the Supreme Court's denial of certiorari, Michigan's attorney general is reportedly considering whether to litigate the dispute in a lower federal court.[83] Environmental advocacy groups, some of which filed amicus briefs in support of Michigan's motion to reopen, may also decide to file separate lawsuits in federal district court.[84]

Canadian Concern

For many decades, the United States and Canada have conducted a major cooperative program to deal with the consequences arising from the introduction of the non-native sea lamprey, *Petromyzon marinus*, to the Great Lakes. Through the Great Lakes Fishery Commission, the governments of the United States and Canada, together with neighboring states and provinces, spend millions of dollars annually to control this invasive parasite and limit its damage to sport and commercial fisheries.

Canada has assessed the risks posed by the introduction of Asian carp,[85] concluding that the risk of impact would be high in some parts of Canada, including the southern Great Lakes basin, by the four species of Asian carp. Canada is currently addressing these concerns through its participation in the bilateral Great Lakes Fishery Commission.

Congressional Interest

Section 126, Title I, of P.L. 111-85 directed the Corps to implement additional measures to prevent aquatic nuisance species from bypassing the Chicago Sanitary and Ship Canal Dispersal Barrier Project and to prevent aquatic nuisance species from dispersing into the Great Lakes. On February 9, 2010, the House Transportation and Infrastructure Subcommittee on Water Resources and Environment held a hearing on Asian carp in the Great Lakes. On February 25, 2010, the Senate Energy and Natural Resources Subcommittee on Water and Power held a hearing to examine the science and policy behind efforts to prevent the introduction of Asian carp into the Great Lakes. In addition, several bills have been introduced in the 111th Congress to address multiple concerns about Asian carp.

- Several bills propose to amend the Lacey Act to add bighead carp (S. 1421/H.R. 3173, H.R. 48, and Section 171 of S. 237) to the list of injurious species that are prohibited from being imported or shipped interstate.[86] The Senate Environment and Public Works Subcommittee on Water and Wildlife held a hearing on S. 1421 on December 3, 2009, and the full committee reported this bill on May 5, 2010 (S.Rept. 111-181).

- H.R. 51 would direct the U.S. Fish and Wildlife Service to study the feasibility of various approaches to eradicating Asian carp from the Great Lakes watershed.
- H.R. 4472 and S. 2946 would direct the Secretary of the Army to take action with respect to the Chicago waterway system to prevent the migration of bighead and silver carp into Lake Michigan, including closing O'Brien and Chicago Locks.
- H.R. 4604 would direct the Secretary of the Army to prevent the spread of Asian carp in the Great Lakes and their tributaries.
- Section 172 of S. 237 would direct the Secretary of the Interior to establish an interbasin and intrabasin monitoring program to monitor the movement of aquatic invasive species in interbasin waterways, assess the efficacy of dispersal barriers and other options for preventing the spread of invasive species, and identify potential sites for dispersal barrier demonstration projects.

End Notes

[1] Information from U.S. Geological Survey Fact Sheet, at http://nas.er.usgs.gov/queries/FactSheet.asp?speciesID=514; and A.J. Mitchell and A.M. Kelly, "The Public Sector Role in the Establishment of Grass Carp in the United States," Fisheries, Vol. 31, no. 3 (March 2006):1 13-121.

[2] F.J. Guscio and E.O. Gangstad, *Research and Planning Conference on the Biological Control of Aquatic Weeds with the White Amur*, prepared for the interagency Research Advisory Committee, Aquatic Plant Control Program, Office of the Chief of Engineers, Department of Army, 1970.

[3] The U.S. Fish and Wildlife Service established a Triploid Grass Carp Inspection Program in 1985 to certify that only genetically triploid (i.e., sterile) grass carp are shipped among 32 states restricting the import of any non-sterile grass carp. For more information on this program, see http://www.fws.gov/policy/aquatichandbook/Volume_9/Volume9.htm.

[4] Information from U.S. Geological Survey Fact Sheet, at http://nas.er.usgs.gov/queries/FactSheet.asp?speciesID=573.

[5] Information from U.S. Geological Survey Fact Sheet, at http://nas.er.usgs.gov/queries/FactSheet.asp?speciesID=549.

[6] W.L. Shelton and R. O. Smitherman, "Exotic Fishes in Warm-Water Aquaculture," *Distribution, Biology, and Management of Exotic Fishes*, W.R. Courtenay, Jr. and J.R. Stauffer, eds., Baltimore, MD: The Johns Hopkins University Press, 1984, p. 262-30 1.

[7] Scott Henderson, *An Evaluation of Filter Feeding Fishes for Removing Excessive Nutrients and Algae from Wastewater*, U.S. Environmental Protection Agency, Project Summary, EPA-600/S2-83-019, May 1983.

[8] Information from U.S. Geological Survey Fact Sheet, at http://nas.er.usgs.gov/queries/FactSheet.asp?speciesID=551.

[9] W.L. Shelton and R. O. Smitherman, "Exotic Fishes in Warm-Water Aquaculture," *Distribution, Biology, and Management of Exotic Fishes*, W.R. Courtenay, Jr. and J.R. Stauffer, eds., Baltimore, MD: The Johns Hopkins University Press, 1984, p. 262-30 1.

[10] At one time, the market for this species produced by aquaculture was primarily the ethnic live-fish trade in large cities. However, live sale of this species is now prohibited in many cities. For California markets, these fish were killed before entering the state to keep them as fresh as possible. An exception is New York City, where it is still legal to sell live bighead carp, but they must be killed before they leave the store.

[11] This species also was raised previously by aquaculture operations in Kansas and Illinois.

[12] Scott Henderson, *An Evaluation of Filter Feeding Fishes for Removing Excessive Nutrients and Algae from Wastewater*, U.S. Environmental Protection Agency, Project Summary, EPA-600/S2-83-019, May 1983.

[13] Duane Chapman, research fisheries biologist, U.S. Geological Survey, Columbia Environmental Research Center, Columbia MO, personal communication, February 26, 2010.

[14] Greg Conover, "The Asian Carp Working Group Update," *ANS Task Force Spring Meeting Minutes*, May 26-27, 2004, p. 35-37; Available at *http://www.anstaskforce.gov/ Minutes/ Spring04_Minutes.pdf*.

[15] For more background on prevention and control methods, see CRS Report RL30 123, *Invasive Non-Native Species: Background and Issues for Congress*, by M. Lynne Corn et al.

[16] Established in 1954 under the bilateral U.S./Canada Convention on Great Lakes Fisheries.

[17] See http://www.glfc.org/fishmgmt/carp

[18] See http://www.asiancarp.org/rapidresponse/documents/AsianCarp.pdf.

[19] See http://www.glfc.org/fishmgmt/carp

[20] Cynthia S. Kolar and David M. Lodge, "Ecological Predictions and Risk Assessment for Alien Fishes in North America," *Science*, vol. 298 (November 8, 2002), pp. 1233-1236.

[21] According to the U.S. Fish and Wildlife Service, Asian carp degrade waterfowl habitat and put waterfowl production areas at risk. Reductions of waterfowl populations could decrease hunting opportunities and associated economic impacts from hunting expenditures.

[22] Dan Brannan, "Business Hopes to Sell Invasive Carp to Asians," *The Telegraph* , March 14, 2010.

[23] This discussion only considers the U.S. economy; Canadian fisheries and recreation might also be affected. See the later section "Canadian Concern."

[24] Social welfare is a measure of the well-being of society or of a community. Estimates of changes in social welfare determine whether society loses or gains from a given action.

[25] U.S. Department of the Interior, Fish and Wildlife Service, and U.S. Department of Commerce, Census Bureau, *2006 National Survey of Fishing, Hunting, and Wildlife-Associated Recreation*, Washington, DC, 2007.

[26] Southwick Associates, *Sportfishing in America: An Economic Engine and Conservation Powerhouse*, American Sportfishing Association, Multistate Conservation Grant Program, 2007. Hereinafter cited as "Southwick Associates 2007."

[27] Total impacts include direct, indirect, and induced impacts as money is cycled through the economy, in this case as a result of expenditures on recreational fishing equipment and trips.

[28] Southwick Associates 2007.

[29] See http://www.glerl.noaa.gov/seagrant/FEE/05-504-Economics.pdf.

[30] In this case, landed value is the amount paid to fishermen at the dock.

[31] U.S. Department of Commerce, National Marine Fisheries Service, *Fisheries of the United States 2008*, Silver Spring, MD, July 2009.

[32] Ronald E. Kinnunen, *Great Lakes Commercial Fisheries*, Michigan Sea Grant Extension, Marquette, MI, August 2003.

[33] Norman S. Baldwin, Robert W. Saafeld, and Maragret A. Ross, et al., *Commercial Fish Production in the Great Lakes 1867-1977*, Great Lakes Fishery Commission, Technical Report No. 3, Ann Arbor, MI, September 1979. Hereinafter cited as Great Lakes Fishery Commission 1979.

[34] Great Lakes Fishery Commission 1979.

[35] Great Lakes Commission, *Great Lakes Recreational Boating's Economic Punch*, Ann Arbor, MI, 2004. Hereinafter cited as "Great Lakes Commission 2004."

[36] Great Lakes Commission 2004.

[37] For example, see John Schneider, "Asian Carp's Threat Goes Far Beyond Economics," *Lansing State Journal*, June 3, 2010; available at *http://www.lansingstatejournal.co m/article/20100603/COLUMNISTS09/6030341.*

[38] Although the Corps has the primary authority to maintain the CAWS for navigation, multiple federal, state, and local entities also possess authorities that must be considered in the context of management actions in the CAWS. Some of these entities include the State of Illinois, the Metropolitan Water and Reclamation District of Greater Chicago, the City of Chicago, the Fish and Wildlife Service, the Environmental Protection Agency, and the U.S. Coast Guard.

[39] Today the I&M Canal remains open as a state park site. The I&M Canal's potential to move Asian carp into other CAWS canals has been an additional item of discussion in recent invasive species debates.

[40] The canal was designed to run southwest from Lake Michigan toward the Mississippi at a small gradient.

[41] The Corps and the MWRD coordinate during severe rainstorms, and may open both the locks themselves and the sluice gates to allow for discharge of floodwaters into Lake Michigan to prevent flooding of downtown Chicago. This last occurred in 2008.

[42] For additional information, see http://www.ndc.iwr.usace.army.mil/lpms/pdf/lpmsstat_v3.pdf.

[43] U.S. Army Corps of Engineers, *Waterborne Commerce Statistics Center*, 2008. http://www.ndc.iwr.usace.army.mil/ lpms/lock2008web.htm.

[44] Ibid. According to Corps statistics, approximately 6.8 million tons in bulk commodities transported through the O'Brien Lock in 2008, while 105,000 tons of commodities were transported through the Chicago Lock in 2008. For additional analysis of vessel movement and lockages based on Corps data, see Joel Brammeier, Irwin Polls, and Scudder Mackey, *Preliminary Feasibility of Ecological Separation of the Mississippi River and the Great Lakes to Prevent the Transfer of Aquatic Invasive Species*, Alliance for the Great Lakes, 2008 Project Completion Report, Chicago, IL, November 2008, pp. 50-55.

[45] See U.S. Army Corps of Engineers, *Waterborne Commerce of the United States, Calendar Year 2008, Part 3— Waterways and Harbors, Great Lakes,* IWR-WCUS-08-03, Alexandria, VA, 2008. Available at *http://www.iwr.usace.army.mil/ndc/wcsc/ pdf/wcusgl08. pdf.*

[46] The study was included as an Appendix to Michigan's recent Supreme Court filing, and is available at *http://www.michigan.gov/documents/ag/1-Appendix-Renewed_Motion_ 310133 _7.pdf.* For more information on this litigation, see page 16 of this chapter.

[47] See http://www.ilchamber.org/lockclosingstudy.html.

[48] Joseph P. Schwieterman, An Analysis of the Economic Effects of Terminating Operations at the Chicago River Controlling Works and the O'Brien Locks On the Chicago Area Waterway System, DePaul University, Chicago, IL, April 7, 2010, at *http://www.ilchamber. org/documents/lockstudy/* DePaul%20University%20Study%20on%20Terminating%20Lock%20Operations.pdf.

[49] 16 U.S.C. § 4701.

[50] The waters of the CAWS were widely noted to be polluted and oxygen-deprived through the early 1980s. These conditions likely prevented the spread of aquatic species through the area over the earlier history of the CAWS. Recent efforts to clean up the waterway have also made possible the survival of many species in the area, including invasive species.

[51] 16 U.S.C. § 4722(i)(3). Although the barrier was authorized and designed to repel multiple aquatic invasive species, the primary goal of the original barrier was to impede the downstream movement of round goby from the Great Lakes to the Mississippi River basin. Because of funding and construction delays, the demonstration barrier was not operational

in time to prevent this movement, and round goby were found downstream of the barrier site in 1999.

52 A full history of the demonstration barrier, including the rationale for the preferred barrier technology, is available at http://www.seagrant.wisc.edu/ais/default.aspx?tabid=1543.

53 The demonstration barrier was originally authorized in the National Invasive Species Act of 1996 (P.L. 104-3 32) and its funding level was increased in Emergency Supplemental Appropriations Act for Defense, the Global War on Terror, and Hurricane Recovery, 2006 (P.L. 109-234). Funding for Barrier II was first provided as an environmental restoration project under WRDA 1986 (P.L. 99-662, §1135) in 2002 and required a local cost sharing partner. The project was subsequently authorized at a level of $9 million in the District of Columbia Appropriations Act, 2005 (P.L. 108-335, § 345). In WRDA 2007 (P.L. 110-114), Congress consolidated the multiple authorizations for barrier construction and authorized the Corps to permanently operate both barriers at a 100% federal cost.

54 Personal Communication with Charles Shea, Dispersal Barrier Project Director, Army Corps of Engineers, Chicago District, February 24, 2010.

55 Office of the Assistant Secretary of the Army, *FY 2011 Civil Works Budget for the U.S. Army Corps of Engineers.*, Washington, DC, February 2010, p. LRD-132.

56 Illinois Department of Natural Resources, *Bighead Asian Carp Found in Chicago Sanitary and Ship Canal*, December 3, 2009. Available at *http://dnr.state.il.us/pubaffairs /2009/ December/asianCarp3Dec2009.htm*. At the time, this finding was significant for its confirmation of Asian carp presence in the CSSC. In June, a fish was discovered upstream of the barriers in Lake Calumet. For more information, see "Monitoring" section.

57 See 121 Stat. 1121. The Corps is studying four areas in this regard: optimal operating parameters for the barriers, ANS barrier bypass, ANS human transfer, and ANS abundance reduction.

58 U.S. Army Corps of Engineers—Chicago District, *Interim I Dispersal Barrier Bypass Risk Reduction Study & Integrated Environmental Assessment*, Final Report, Chicago, IL, January 2010. Available at http://www.lrc.usace.army.mil/pao/ANS _DispersalBarrier EfficacyStudy _Interim _I_FINAL.pdf.

59 U.S. Army Corps of Engineers—Chicago District, Interim III Dispersal Barrier Efficacy Study: Modified Structures and Operations, Illinois and Chicago Area Waterways, Risk Reduction Study and Integrated Environmental Assessment, Final Report, Chicago, IL, June 2, 2010. p iii. Available at http://www.lrc.usace.army.mil/pao/02June20 10_InterimIII.pdf. Notably, the Corps did not consider extended lock closure (i.e., more than two months) under this study.

60 U.S. Army Corps of Engineers—Chicago District, Interim IIIA Dispersal Barrier Efficacy Study: Modified Fish Dispersal Deterrents, Illinois and Chicago Waterways Risk Reduction Study and Integrated Environmental Assessment, Chicago, IL, April 2010. Available at http://www.lrc.usace.army.mil/pao/02June2010_InterimIIIA.pdf.

61 U.S. Congress, House Committee on Transportation and Infrastructure, Subcommittee on Water Resources and Environment, *Statement of Professor David Lodge, Director, Center for Aquatic Conservation,* hearing on Asian Carp and the Great Lakes, 111th Cong., 2nd sess., February 8, 2010. Appendix: Laboratory Audit Report, Lodge Laboratory, Department of Biological Sciences, University of Notre Dame.

62 See http://www.asiancarp.org/documents/FINALRCCCarpDiscoveryLakeCalumet June232010. pdf.

63 See Leo G. Nico and J. D. Williams, *Black Carp: A Biological Synopsis and Updated Risk Assessment*, U.S. Geological Survey, Final Report to the Risk Assessment and Management Committee of the ANSTF., Gainesville, FL, 2001, available at http://www.fisheries and C. S. Kolar, D. C. Chapman, and W. R. Courtenay et al., *Asian Carps of the Genus Hypophthalmichthys (Pisces, Cyprinidae):A Biological Synopsis and Environmental Risk Assessment*, U.S. Geological Survey, Report to the Fish and Wildlife Service, LaCrosse,

WI, 2005, available at *http://www.fws.gov/contaminants/OtherDocuments/* ACBSRA Final Report2005.pdf.

[64] The Lacey Act, 16 U.S.C. §§ 3371-3378, makes it unlawful to import, export, sell, acquire, or purchase fish, wildlife or plants taken, possessed, transported, or sold (1) in violation of U.S. or Indian law or (2) in interstate or foreign commerce involving any fish, wildlife, or plants taken, possessed or sold in violation of state or foreign law. Under this law, designated injurious species are identified at 50 C.F.R. § 16. See also *http://www .anstask force.gov/Documents/* Injurious_Wildlife_Fact_Sheet_2007.pdf.

[65] Greg Conover, Rob Simmonds, and Michelle Whalen, *Management and Control Plan for Bighead, Black, Grass, and Silver Carps in the United States,* Aquatic Nuisance Species Task Force, Asian Carp Working Group, Washington, DC, November 2007.

[66] Asian Carp Workgroup, *Asian Carp Control Strategy Framework*, Washington, DC, May 5, 2010, p. vi. Available at http://www.asiancarp.org/Documents/AsianCarpControlStrategyFrameworkMay2010.pdf.

[67] P.L. 110-114, § 345.

[68] Brammeier et al., p. 99. The study recommended that the Corps take a more comprehensive look at the available engineering options for ecological separation.

[69] See http://www.asiancarp.org/Documents/AsianCarpControlStrategyFrameworkMay2010.pdf.

[70] For a complete summary of each recommendation, including funding sources, see *Asian Carp Control Strategy Framework*, ibid., pages 13-33, or Table 1 in that report.

[71] The United States Supreme Court has "original and exclusive jurisdiction of all controversies between two or more States." 28 U.S.C. § 1251(a).

[72] When the Supreme Court exercises its original jurisdiction, disputes between states are filed under "Original" dockets that are assigned a number. The state of Michigan argued that the Court had original jurisdiction over this dispute based on a decree issued in 1967, which resolved a dispute between the Great Lakes states and Illinois over the amount of water Illinois could withdraw from Lake Michigan for sanitary and navigational purposes. See *Wisconsin v. Illinois*, 388 U.S. 426 (1967). The Court retained original jurisdiction over the dispute. *Id.* Accordingly, docket nos. 1, 2, and 3, Original, remain "open" for the purpose of resolving additional disputes between Great Lakes states and Illinois, provided that such disputes relate to the operation of the Chicago Area Waterway System, which links the Mississippi River to Lake Michigan. Michigan's motion to reopen docket nos. 1, 2, and 3, Original, is available at http://www.supremecourt.gov/SpecMastRpt/Orig%201,%202%20&%203%20Motion%20to%20Reopen.pdf.

[73] The state of Illinois is a party to this dispute because, according to the state of Michigan, Illinois is ultimately responsible for the operation of the Chicago Waterway System, which is jointly operated by the Army Corps of Engineers and Metropolitan Water Reclamation District of Greater Chicago. *Mich. Mot. for Prelim. Injunction.* In response, the state of Illinois and the United States argue that Illinois is not a proper party to this dispute because the state does not directly exercise day-to-day control over the operation of the Chicago Waterway System. *Ill. Response; U.S. Response*. The parties' filings are publicly available at http://www.supremecourt.gov/SpecMastRpt/ RecentFilingsinOriginalNos_1_2_3.aspx (links to PDF versions of filings).

[74] Minnesota, New York, Ohio, Wisconsin, and the Canadian Province of Ontario filed briefs in support of Michigan's request for a preliminary injunction.

[75] Neither the state of Illinois nor the United States denied the threat posed to the Great Lakes by the spread of Asian Carp in their respective responses to Michigan's request for a preliminary injunction, but rather argued that the requested relief is unnecessary in light of current efforts to prevent the spread of Asian carp into the Great Lakes through the Chicago Waterway System.

[76] The Supreme Court declined to address the merits of Michigan's arguments and simply denied Michigan's motion for a preliminary injunction in a one-sentence order. See *http://*

*www.supremecourtus.gov/orders/courtorders/*01 1910zor.pdf (order denying Michigan's request for a preliminary injunction at page 3).

[77] See http://www.supremecourt.gov/SpecMastRpt/1-Renewed%20Motion%20for%20PI.pdf (Michigan's renewed motion for preliminary injunction).

[78] See http://www.supremecourt.gov/orders/courtorders/032210zor.pdf (order denying Michigan's renewed request for a preliminary injunction at page 2).

[79] *Mich. Mot. to Reopen* at 29 (available at *http://www.supremecourt.gov/ SpecMast Rpt/ Orig%201,%202%20&* %203%20Motion%20to%20Reopen.pdf).

[80] *Id.* at 29-30.

[81] See http://www.supremecourt.gov/orders/courtorders/042610zor.pdf (order denying Michigan's motion to reopen docket numbers 1, 2, and 3, Original at pages 1-2).

[82] *Id.*

[83] See Gabriel Nelson, "Great Lakes: Supreme Court Halts Asian Carp Lawsuit," *Greenwire*, April 26, 2010, available at http://www.eenews.net/Greenwire/2010/04/26/4/.

[84] *Id.*

[85] Available at http://www.dfo-mpo.gc.ca/csas/Csas/DocREC/2004/RES2004_103_E.pdf.

[86] FWS has considered listing bighead carp as an injurious species under the Lacey Act, but decided not to take this action because of the negative economic impact such a listing might have had upon the U.S. aquaculture industry. See 68 *Federal Register* 54409, September 17, 2003.

In: Troubled Waters: Asian Carp and the... ISBN: 978-1-61761-766-9
Editors: Deborah M. Maeda

Chapter 2

ASIAN CARP AND THE GREAT LAKES HEARING, BEFORE THE SUBCOMMITTEE ON WATER RESOURCES AND ENVIRONMENT

SUMMARY OF SUBJECT MATTER

TO:	Members of the Subcommittee on Water Resources and Environment
FROM:	Subcommittee on Water Resources and Environment Staff
SUBJECT:	Hearing on the "Asian Carp and the Great Lakes"

PURPOSE OF THE HEARING

The Subcommittee on Water Resources and Environment will meet on Tuesday, February 9, 2010, at 2:00 p.m., in room 2167 of the Rayburn House Office Building to hear testimony from representatives of the U.S. Environmental Protection Agency (EPA), the U.S. Army Corps of Engineers (Corps), the States of Illinois and Michigan, academia, and interested stakeholders on the issue of preventing the induction of the aquatic invasive species, the Asian carp, into the Great Lakes.

BACKGROUND

I. Asian Carp Introduction and Threat to Aquatic Ecosystems

Two species of carp – the bighead carp and silver carp – hereinafter referred to as Asian carp, were first introduced into the southern United States in the early 1970s to remove algae from farmed catfish ponds. Flooding in the early 1990s caused many of the catfish farm ponds to overflow, and Asian carp were released into local waterways in the Mississippi River basin. The carp have since migrated northward up the Mississippi River, becoming the most abundant species in some areas of the River.[1]

Three characteristics of Asian carp have complicated efforts to control the northward migration of the species towards the Great Lakes including its: (1) large size (up to four feet in length and 100 pounds); (2) rapid rate of reproduction (spawning carp can produce between 50,000 to 1 million eggs); and (3) aggressive nature (Asian carp are easily agitated by loud noises, such as boat engines, and have been documented jumping out of rivers into ongoing boaters).

Asian carp are of particular concern to the Great Lakes region because of the potential harm they could cause to the native ecosystem. Asian carp currently present in the Mississippi River physically lack a stomach, so they must continually feed on aquatic vegetation. Native fish species have difficulty competing with Asian carp because of their rapid consumption of resources.

Additionally, Asian carp can readily adapt to varying environmental conditions, exacerbating the ecosystem concerns for the Great Lakes. According to the U.S. Fish and Wildlife Service, "bighead and silver carp all prefer a wide temperature range, indicating their ability to thrive from the northernmost waters of the Great Lakes to the waters of the middle Mississippi River Basin. . . . Asian carp species that have become adapted to life in the Great Lakes would also likely invade the Lakes' tributary streams and rivers where they would most likely spawn."[2]

II. The Chicago Sanitary and Ship Canal[3]

The Chicago Area Waterway System is a system of canals and natural waterways that serves as both a navigation link between Lake Michigan and the Mississippi River system and as an outlet for the stormwater and effluent from the City of Chicago. The canal system extends between Lake Michigan and the Des Plaines River, a tributary of the Illinois River and ultimately of the Mississippi River. The canal system was originally constructed to permit Chicago to dilute and dispose of its wastewater without allowing it to enter Lake Michigan. Using the canal system, Illinois redirected the Chicago River, which naturally flowed east into Lake Michigan, to flow west, carried by the canal system into the Des Plaines River. The Chicago River Controlling Works were constructed at the confluence of the Chicago River and Lake Michigan.

The permanent connection between Lake Michigan and the Mississippi River drainage basins was finalized with the completion of the Chicago Sanitary and Ship Canal (Canal) in 1900. Subsequent construction included the dredging and reversal of the Calumet River, the erection of the Thomas J. O'Brien Lock and Dam on that river, and the construction of the Cal-Sag Channel linking the Calumet with the main canal.

By statute, the Corps operates and maintains the Canal as necessary to sustain navigation from Chicago Harbor on Lake Michigan to Lockport on the Des Plaines River. Vessels enter and exit the Chicago end of the canal system through the O'Brien Lock and through the lock facilities at the Chicago River Controlling Works (the Chicago Lock). The Corps owns both locks and operates them in accordance with applicable regulations and agreements with the Metropolitan Water Reclamation District of Greater Chicago (Water District).

Most commercial boat traffic between Lake Michigan and the canal system now passes through the O'Brien Lock, including barge traffic recently rerouted from the Chicago Lock. About seven million tons of cargo[4] pass through the O'Brien Lock each year, as do 18,000 recreational boats. Additional cargo, ferry, and pleasure boats use the Chicago Lock. The locks are also used by the U.S. Coast Guard stations on the Lake Michigan side of the locks in responding to safety emergencies on the canal and in patrolling critical infrastructure facilities in the river system.

In addition, both the Chicago River Controlling Works and the O'Brien Lock are used for flood control purposes. Both facilities include sluice gates connected to the locks, which are used to combat the risk of flooding during significant rainstorms by drawing water from the canal system into Lake Michigan. The Corps owns the sluice gates at the O'Brien Lock and operates them under the direction of the Water District. The Water District owns and operates the sluice gates at the Chicago River Controlling Works. In very severe rainstorms, in addition to opening the sluice gates, the Water District requests the Corps to open the Chicago and O'Brien lock gates as well, to permit additional water to be diverted into Lake Michigan. Both locks were last opened for this flood control purpose in September 2008.

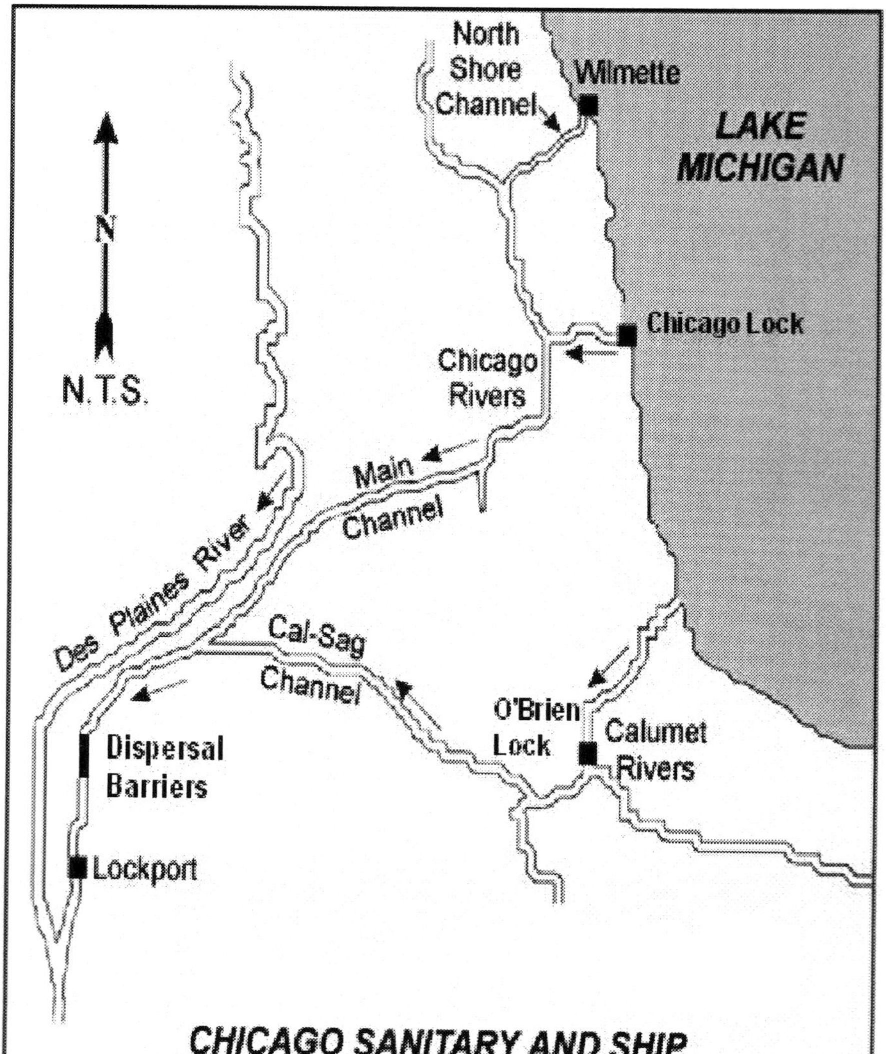

Source: U.S. Army Corps of Engineers, Chicago District

III. Federal Efforts to Combat Asian Carp

A. Electric Dispersal Barrier

Congress first recognized the potential for invasive species, including Asian carp, to enter the Great Lakes via the Canal in 1996. In that year, Congress included language in the National Invasive Species Act of 1996 (16 U.S.C. § 4713, *et seq.*), directing the Corps to study and implement a dispersal system along the Canal to prevent the introduction of Asian carp into the Great Lakes.

Pursuant to this authority, the Corps elected to construct an electric barrier system because it was, generally, non-lethal to fish and does not interfere with navigation.[5]

In April 2002, the Corps began operation of the first electric barrier in Romeoville, IL (now known as Barrier I) as a demonstration of the technology to deter invasive species. The system consists of steel cables that are secured to the bottom of the Canal. A low-voltage, pulsing direct current (DC) is sent through the cables, creating an electric field in the water. This field is uncomfortable for fish and most do not swim through it. Barrier I was intended to only operate for a few years as a demonstration project. After Barrier I was constructed, further research by the Corps suggested that smaller or juvenile fish required more voltage beyond the capacity of Barrier I.

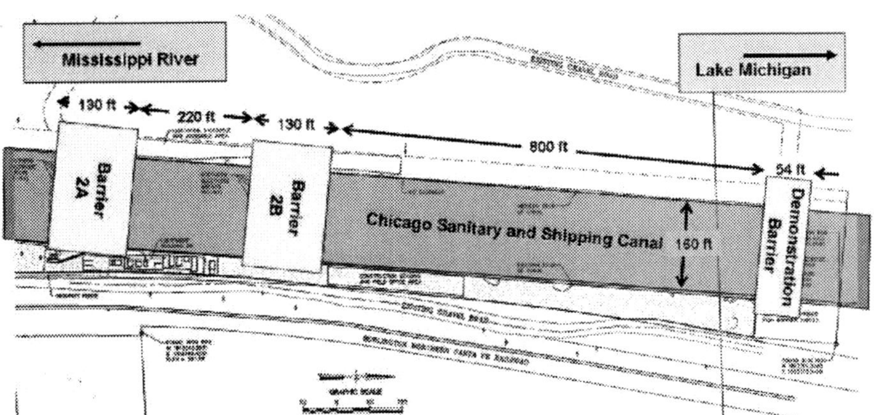

Source: U.S. Army Corps of Engineers, Chicago District

In 2003, the Corps initially approved the construction of a permanent barrier – Barrier IIA – under its continuing authorities program (section 1135 of the Water Resources Development Act of 1135). This project, which was later specifically authorized by Congress through the District of Columbia Appropriations Act (2005), includes several design improvements identified during the study of Barrier I. Barrier II is able to generate a stronger electric field over a larger area and has two control houses, Barrier IIA and IIB. Each control house can be operated independently, but the Corps' goal is to have both operated simultaneously. In the Water Resources Development Act of 2007, Congress authorized the Corps to complete Barrier II, upgrade Barrier I and make it permanent, and to operate the barrier system at Federal expense.

Barrier IIA was completed in 2006 and was fully operational in April 2009. Barrier IIB is currently under construction, utilizing funds from the American Recovery and Reinvestment Act (P.L. 111-5), and is expected to be complete by the end of 2010.

B. Efficacy Study

While the Canal barriers provide a level of protection against the direct migration of Asian carp through the Canal to the Great Lakes, these species potentially could bypass the barriers during a flooding event and cross over into the adjacent Des Plaines River or Illinois and Michigan Canal. To minimize this potential, Congress authorized[6] the Corps to conduct an efficacy study "to prevent aquatic nuisance species from bypassing the Canal Dispersal Barrier Project" and entering the Great Lakes.

On January 10, 2009, the Corps released its Final Report with recommendations for interim risk reduction measures resulting from the efficacy study. The recommended plan is for the placement of 34,600 feet of concrete barricades and 33,400 feet of chain link fencing (with 1/4" openings) along 68,000 feet (~13 miles) of flood prone areas of the Canal upstream of the barriers. The total cost of these interim measures is $13,174,000, which was provided through the Great Lakes Restoration Initiative.[7] The Corps expects to award a contract for these measures in March 2010, and expects to complete construction of the interim measures by October 2010.

C. Great Lakes Mississippi River Interbasin Feasibility Study

The Corps is also carrying out a feasibility study, pursuant to section 3061 of the Water Resources Development Act of 2007, to investigate long-term efforts for preventing invasive species from entering into the Great Lakes.

The initial scoping of the Interbasin Study will examine several potential pathways for movement of aquatic invasive species into the Great Lakes, including the Canal and connected waterways, the Ohio River and Lake Erie, and the Fox and Wisconsin Rivers. Although the scope of the study covers multiple pathways into Great Lakes, the Corps expects to prioritize the feasibility study for the Canal, expected to be completed within 2 1/2 years. The remainder of the study is expected to be completed within five years. The study has a multi-year timeframe, but the Corps expects individual decisions can be made as the appropriate section of the study is finalized, rather than waiting for the entire study to conclude. According to the Corps, approximately $450,000 of fiscal year 2009 and 2010 appropriations is available for the Interbasin study.

D. eDNA Sampling Technique

In August 2009, the Corps entered into an agreement with the University of Notre Dame to use an experimental technique called environmental DNA (eDNA) testing to identify Asian carp in the Canal. This technique involves water sampling, filtering for organic solids, and analyzing the DNA for genetic markers specific to the Asian carp.

The initial results of the eDNA technique found Asian carp eDNA south of the barrier system only. More samples were taken further upstream of the barriers. Four samples from the Calumet-Sag Channel area revealed Asian carp eDNA. Subsequent tests have been negative, although this may have been due to colder winter temperatures. Sampling is still being conducted to test for the presence of Asian carp north of the barrier system.

E. Ballast and Bilge Water Restrictions

Water taken on by vessels for stability (ballast) or water that accumulates in void spaces of vessels (bilge) can contain invasive species. In September 2009, the barge industry agreed to cease ballasting on either side of the barrier, at the Coast Guard's request.

IV. State Efforts to Combat the Asian Carp

The Illinois Department of Natural Resources (IDNR) is the lead responding agency member of the Asian Carp Rapid Response Workgroup. The Workgroup's other members are the Coast Guard, the Metropolitan Water Reclamation District of Greater Chicago, the Corps, the EPA, and the United States Fish and Wildlife Service. The Workgroup has conducted several recent efforts to identify and kill Asian carp in the Canal.

A. Rotenone Application

In December 2009, Barrier IIA was taken offline for routine maintenance. To prevent the passage of Asian carp through the barrier system during maintenance, IDNR applied the fish poison (Rotenone) to a 5.7-mile portion of the Canal south of the barrier system. Other States involved in the Rotenone application included Michigan, Indiana, Wisconsin, and Canada, which provided personnel and equipment. The Rotenone killed approximately 30,000 fish in the Canal. One Bighead carp was found south (riverward) of the barrier system.

B. Fish Netting and Tagging

In response to the eDNA findings along the Calumet-Sag Channel, Illinois contracted with commercial fisherman experienced in catching Asian carp. From December 1-7, 2009, over 1,000 fish were caught and identified, but no Asian carp were identified.

PENDING LEGISLATION

Several legislative proposals have been introduced in the House of Representatives to address the issue of the Asian carp.

H.R. 51, the "Eradicating Asian Carp in the Great Lakes Study Act," (introduced by Representative Mark Kirk of Illinois) requires the Director of the United States Fish and Wildlife Service to study and report to Congress on methods for eradicating Asian carp from the Illinois Waterway System, including harvesting them from the Illinois River, and repopulating the River with native species. H.R. 51 was referred to the Committee on Natural Resources.

H.R. 3173, the "Asian Carp Prevention and Control Act", (introduced by Representative Judy Biggert of Illinois) amends the Lacey Act to add the bighead carp to the list of injurious species that are prohibited from being shipped or imported into the United States. H.R. 3173 was referred to the Committee on the Judiciary.

H.R. 4472, the "Close All Routes and Prevent Asian Carp Today Act of 2010," (introduced by Representative Dave Camp of Michigan) directs the Secretary of the Army to immediately close and cease operation of the O'Brien and Chicago locks to prevent the migration of bighead and silver carps into Lake Michigan. H.R. 4472 was referred to the Committee on Transportation and Infrastructure.

AGENDA

Panel I

Mr. Cameron Davis
Senior Adviser to the Administrator
United States Environmental Protection Agency

Major General John W. Peabody
Commander, the Great Lakes and Ohio River Division
United States Army Corps of Engineers

Director Marc Miller
Illinois Department of Natural Resources

Director Rebecca Humphries
Michigan Department of Natural Resources and Environment

Secretary Matt Frank
Wisconsin Department of Natural Resources

Panel II

Professor David Lodge
Director, Center for Aquatic Conservation
Professor of Biological Sciences
University of Notre Dame

Dr. Michael Hansen
Chair, Great Lakes Fishery Commission

Mr. Del Wilkins
Vice President of Terminal Operations & Business Development

Canal Barge Company, Inc.

Testifying on behalf of The American Waterways Operators

Mr.Joel Brammeier
President and CEO
Alliance for the Great Lakes

End Notes

[1] EPA, Asian Carp and the Great Lakes (November 25, 2009), *http://www.* epa.gov/glnpo/invasive/asiancarp/.

[2] *See* Declaration of Charles M. Wooley, Deputy Regional Director of the Midwest Region, United States Fish and Wildlife Service, *Wisconsin v. Illinois and Metropolitan Sanitary District of Greater Chicago* (Jan. 4, 2010).

[3] *Source:* Memorandum for the United States in Opposition, *Wisconsin v. Illinois and Metropolitan Sanitary District of Greater Chicago* (Jan. 2010).

[4] Predominately coal and other fossil fuels, bulk construction materials (e.g., sand and gravel), iron and steel products, and chemical products.

[5] Corps, Chicago Sanitary and Ship Canal Aquatic Nuisance Species Dispersal Barriers (August 27, 2009), http://www.lrc.usace.army.mil/projects/fish_barrier/index.html.

[6] Section 126 of the Energy and Water Development and Related Agencies Appropriations Act of 2010, Pub. L. 111-85 (2009).

[7] Department of the Interior, Environment, and Related Agencies Appropriations Act, 2010, Pub. L. No. 111-88 (2009).

In: Troubled Waters: Asian Carp and the… ISBN: 978-1-61761-766-9
Editors: Deborah M. Maeda

Chapter 3

TESTIMONY OF CAMERON DAVID, SENIOR ADVISOR TO THE ADMINISTRATOR, U.S. ENVIRONMENTAL PROTECTION AGENCY, BEFORE THE SUBCOMMITTEE ON WATER RESOURCES AND ENVIRONMENT, HEARING ON "ASIAN CARP AND THE GREAT LAKES"

Chairwoman Johnson and members of the subcommittee, on behalf of U.S. Environmental Protection Agency Administrator Jackson, thank you for the opportunity to provide the agency's perspective on efforts to prevent the establishment of Asian carp in the Great Lakes.

The administration continues to make restoration and protection of the Great Lakes a national priority as evidenced by President Obama's, significant investment in the ecosystem under his Great Lakes Restoration Initiative (Initiative).

EPA understands the extreme level of concern that the public feels for the ecosystem, their safety while recreating, and for their jobs. We also have an urgent need to keep Asian carp from becoming established in the Great Lakes. As we move forward, we are working to keep Asian carp from becoming established in Lake Michigan. But to do that will require a coordinated, cooperative approach.

I will address EPA's role in efforts in the recent past and multi-stakeholder plans for moving forward.

EPA's Role

First, EPA is tasked with coordinating federal Great Lakes protection and restoration policies and efforts, under Clean Water Act Section 118 and Executive Order 13340. EPA has been doing this and will facilitate the integration of efforts by participating agencies and stakeholders. One of the best weapons we have against Asian carp is a coordinated, cooperative approach through which each agency remains accountable for the work under its authorities, in order to ensure the most effective efforts possible. We will fail if we inhibit such accountability and integration.

This team approach has been successful and will continue to be successful. Specifically, it worked this past December, when EPA helped coordinate an impressive multi-jurisdictional team with Illinois Department of Natural Resources to undertake a "rapid response" action. The action was needed to ensure Asian carp did not move past the U.S. Army Corps of Engineers' (Corps) electric barriers designed to prevent Asian carp migration through the Chicago Sanitary and Ship Canal. Federal, Canadian, municipal, provincial, binational and state agencies – all of whom provided people, funding or equipment – contributed to an effort that, by all accounts, was highly successful despite numerous obstacles. This team approach also led to the draft *Framework* released this week. I will talk more about that in a minute.

The second role that EPA has to play is that of funding. Nearly a year ago, President Obama proposed and, thanks to your help, Congress passed the Great Lakes Restoration Initiative, an unprecedented investment for rehabilitating the nation's largest fresh surface water ecosystem.

EPA is stepping up to use its funding authority under the Initiative to help slow the migration of Asian carp. In December, EPA announced more than $13 million in funding for emergency measures and additional monitoring to be executed by the Corps. This work will address potential bypasses between either the Des Plaines River or the Illinois and Michigan Canal and the Chicago Sanitary Ship Canal during high water events. Currently, we are working with the other federal agencies and Illinois to fast-track additional investments under the Initiative that will address Asian carp populations that may be upstream of the electric barriers.

NEXT STEPS

Using the coordinated team approach, participating agencies have published a draft *Asian Carp Control Strategy Framework* this week. We want to accomplish several things with this *Framework.*

First, we want to provide direction, without restricting ideas and initiatives. One thing we have learned is that the Asian carp situation continues to evolve. Therefore, the *Framework* must unify everyone to generally move in the same direction, while allowing all agencies' responses to adapt as necessary.

Second, we want to establish a multi-tiered defense. This would include structural, chemical, biological, operational and management solutions. We cannot fight biology with engineering alone.

Third, we want to create space for every player – agencies, conservation interests, businesses, the navigation industry, recreational groups and even individuals – to engage in the effort to keep Asian carp from becoming established in the Great Lakes. In short, this *Framework* belongs to everyone.

The Great Lakes region must unite in this effort. The December rapid response action illustrates just what we can achieve when we work together. The *Framework* is not intended to be final; it is intended to be continually improved. The first step is for everyone to have a hand in its development and its execution.

I want to thank you and the Great Lakes Congressional delegation for your concern and support of control efforts. Administrator Jackson, our partner agencies, the states and delegation all share one overriding imperative: to ensure we leave the Great Lakes better for the next generation than the way in which we inherited them.

In: Troubled Waters: Asian Carp and the... ISBN: 978-1-61761-766-9
Editors: Deborah M. Maeda

Chapter 4

Statement of Major General John Peabody, Commander, Great Lakes and Ohio River Division, U.S. Army Corps of Engineers, before the Subcommittee on Water Resources and Environment, Hearing on "Asian Carp and the Great Lakes"

Madam Chair and members of the Subcommittee, I am Major General John Peabody, Commander of the Great Lakes and Ohio River Division, U.S. Army Corps of Engineers. Thank you for the opportunity to testify about ongoing efforts to address the risk to the Great Lakes posed by the migration of two species of Asian carp, the silver and bighead, through the Chicago Area Waterway System. Asian carp represent a grave threat to the Great Lakes fisheries and aquatic resources, including those managed by the National Park Service in areas such as Sleeping Bear Dunes National Lakeshore, Isle Royale National Park and Apostle Islands National Lakeshore. The Army Corps of Engineers (Corps) remains committed to use all available authorities, capabilities, and resources to combat this invasive species. The Corps cannot do this on its own, but continues to work intensively with, and leverage the full capabilities of Federal, State, provincial, bi-national, and municipal agency partners. I would like to briefly describe the Corps of Engineers' role in this important effort, current actions that the Corps is taking, plans for the immediate future and near term, as well as the Corps' longer term strategy.

The Corps' principal role in this effort has been to address potential migration of Asian carp via the most direct pathway, the Chicago Sanitary and Ship Canal, by building, operating, and improving the electrical dispersal barrier system at Romeoville, Illinois. This fish barrier is the largest fielded operational electrical dispersal barrier in the world and constitutes a complex and dynamic project with significant research and development components. It currently consists of two separate barriers, Barrier 1 and Barrier 2A, with a third, Barrier 2B, under construction. Barrier 1 was built as a demonstration barrier for the purpose of preventing migration of the aquatic nuisance species from Lake Michigan into the Chicago Area Waterway System and has limited operational parameters. It has been operating at its design capacity since 2002. Today, Barrier 2A currently stands as the primary impediment to Asian carp migration. Corps of Engineers' laboratory testing shows that the operating parameters we are applying at Barrier 2A are effective at repelling Asian carp. However, we continue to research the optimal operating parameters and will make adjustments to Barrier 2A as research indicates. Any changes will be preceded by a thorough safety review in conjunction with the U.S. Coast Guard. American Recovery and Reinvestment Act funds have accelerated construction of Barrier 2B by a year; construction is now scheduled to be completed this fall. This barrier will ensure redundancy in the system for maintenance and unexpected actions.

The electrical barriers must be turned off periodically for maintenance. The barrier maintenance, which took place in early December, was successfully completed thanks to the synchronized application of rotenone by the Illinois Department of Natural Resources and the cooperation of multiple federal, state, and provincial entities. Application of rotenone ensured that Asian carp would not pass through the barrier during the brief maintenance period. The single Asian carp discovered following this operation was found below the barrier system. This is an outstanding example of the multi-agency collaborative efforts necessary to be successful in this challenging effort.

It is important to recognize that the electrical barriers do not provide a guarantee that Asian carp will be prevented from entering Lake Michigan. Essential to the Corps' operation of the barriers are actions to ensure their efficacy and to address possible bypasses by Asian carp. Section 3061 of the Water Resources and Development Act of 2007 authorizes the Corps to carry out an "Efficacy Study" in order to develop recommendations for permanent solutions to Asian carp bypass scenarios as well as other potential barriers and impediments to Asian carp migration in the Chicago area. Recent authority, Section 126 of the 2010 Energy and Water Development Appropriations Act, allows the Secretary of the Army to approve measures recommended in the Efficacy Study and other emergency measures as necessary until October 28, 2010 – one year after passage of the bill.

Under these authorities, the Corps has already taken action to address possible bypass of the electrical barriers. On January 12, 2010, the Assistant Secretary of the Army (Civil Works) approved an interim report that recommended emergency measures to address potential bypasses of the barriers. The Corps plans to begin construction of those emergency measures this spring. The recommended structural solutions include installing specially designed fence and concrete barriers at locations along the Des Plaines River adjacent to the Chicago Sanitary and Ship Canal where bypass of the fish barrier could occur during flood events, such as the flooding that most recently occurred in September 2008. The Corps also plans to block flow through the Illinois and Michigan Canal at the natural flow divide to prevent circumvention of the barrier via that pathway. The current schedule provides that construction will be completed by this fall. The Final Efficacy Study will evaluate other potential measures of improving the efficacy of the fish barrier in the Chicago Area Waterway System, such as additional electrical barriers or other types of behavioral barriers, controlling ballast water and other potential transit pathways, modified lock operations, and Asian carp population control. The study will be completed this fall. These critical activities are being funded through FY 2010 appropriations provided to the U.S. Environmental Protection Agency (EPA) for the interagency Great Lakes Restoration Initiative. In support of the Initiative, which includes invasive species prevention as one of its highest priorities, EPA allocated $13.5 million to the Corps for implementation of the recommended emergency measures and additional monitoring.

Monitoring Asian carp migration is an essential part of the inter-agency effort. As part of a comprehensive review of the Corps' activities begun in the Fall of 2008, we determined that the tools available at that time, principally netting and electro-fishing conducted primarily by our partner agencies, could tell us the locations where fish were likely located in abundance, but not necessarily how far they had migrated up the system in smaller numbers. As a result, the Corps canvassed the scientific community for alternative methods, and discovered the University of Notre Dame's environmental DNA (eDNA) research in May of 2009. The University's Dr. David Lodge, along with his team and partners from The Nature Conservancy, agreed to apply their emerging technology to assist us in our efforts to improve understanding of where Asian carp may be located.

Results of eDNA research are an important tool in our ability to confront this threat. Asian carp eDNA testing remains to be fully validated and results should be considered preliminary. Further, the identification of eDNA in a waterway cannot, at present, tell us such things as the size of any Asian carp population, how recently Asian carp have been there, whether the DNA came from a living or dead fish, or whether Asian carp tissue or DNA might have been transported in ballast or bilge water, or via other mechanisms, from some remote location. Because eDNA is a new approach to assessing the presence of Asian carp and is being applied operationally before standard independent scientific review could occur, the Corps continues to collaborate with the University of Notre Dame to determine what eDNA does and does not tell us and continues to research how to improve the usefulness of this technology to inform management decisions. This effort is consistent with the Corps' policy of ensuring that its technical, engineering, and scientific work undergoes an open, dynamic, and vigorous review process to ensure confidence in our decisions and policy recommendations – especially when those decisions may have dramatic consequences. While eDNA is important to the overall effort it is important to recognize that it is a technique for improving our ability to detect the presence and map the distribution of Asian carp and not by itself a tool for prevention.

On January 27, 2010, the Corps received a summary Audit Report on the eDNA scientific process and the reliability of the testing and surveillance methods from the U.S. Environmental Protection Agency. Although the report made recommendations for future method development and quality system improvement, the U.S. Environmental Protection Agency auditors expressed overall "confidence in the reliability of the eDNA protocol implemented" by the University of Notre Dame laboratory. The Corps looks forward to intensely reviewing the final Audit Report to inform our judgments and additional research that may be needed.

Sampling by Dr. Lodge's team has returned positive results for DNA from silver and bighead Asian carp, the two species of concern, in various locations. Identification of Asian carp DNA in the Brandon Road pool in August 2009, which is just over 6 miles downstream of the electrical barriers, triggered the Corps' decision to increase the operating parameters of Barrier 2A. Between August and November 2009, all returned results for Asian carp sampled above the fish barrier were negative. On November 17, 2009 Asian carp DNA was reported as having been detected in the Cal-Sag Channel and Calumet River near the O'Brien Lock, in three areas ranging from 10 to 30 miles upstream of the fish barrier. An intensive fishing effort followed and although over 1,000 fish were caught near the O'Brien Lock, none of them were Asian carp. Recently, positive detections of Asian carp eDNA have also been reported north of the fish barrier near the Wilmette Pumping Station and lakeward of the O'Brien Lock. The Corps continues to consult with and rely upon the assessment of our partner agencies, including the U.S. Fish and Wildlife Service, Illinois Department of Natural Resources, and the U.S. Environmental Protection Agency in evaluating these results.

In the meantime, the Corps of Engineers, working through the Asian Carp Regional Coordination Committee, is urgently developing additional measures to apply in the Chicago Area Waterway System once warmer weather in the spring prompts increased fish activity. Measures the Corps is discussing with our partner agencies include modified operations at existing locks and controlling works, installing other types of barriers near the locks, assessing options to block the alternate pathways of the Grand and Little Calumet Rivers, and supporting efforts to reduce or eliminate Asian carp populations that may be present. Such supporting efforts may include intensive fishing efforts around the navigation structures and the application of rotenone by partner agencies. All measures under consideration will be evaluated taking into account stakeholders' use of these structures, especially by public health, security and safety agencies, such as the Chicago Fire and Police Departments, and the Department of Homeland Security, as well as for purposes of flood risk management, navigation, and water quality. To be effective, any measures would have to be taken in concert with actions by other agencies on matters within their expertise and authority to eliminate or reduce the numbers of any Asian carp that may be in the vicinity. The National Invasive Species Council (NISC) was established by Executive Order 13112 to coordinate federal actions on invasive species. The Secretaries of the Interior, Agriculture, and Commerce are the co-chairs of NISC. The Secretary of Defense is a member of NISC. The Corps of Engineers works with its NISC partner agencies to stop the spread of Asian carp and other invasive species.

Finally, building on all of these efforts, the Corps has a long term strategy. The Corps is undertaking the Congressionally-authorized Great Lakes and Mississippi River Interbasin Study to explore the options and technologies that could be applied to reduce the risk of aquatic invasive species transfer throughout multiple points between the Great Lakes and Mississippi River basins. This study will include the possibility of ecosystem separation and will analyze the impact that alternative possible plans would have on the current uses of the Chicago Area Waterway System, including the Chicago Sanitary Ship Canal. The Corps plans to conduct this study in close coordination with partner governmental agencies, such as the U.S. Environmental Protection Agency, the U.S. Fish and Wildlife Service and the U.S. Coast Guard. The initial area of emphasis will focus on the Chicago Area Waterway System and is anticipated to be completed as an interim report before the final study is complete.

In conclusion, the fish barrier remains our most important immediate defense mechanism against Asian carp migration, but the Corps is working with our partner agencies to transition to a multi-tiered defense. Efforts to prevent Asian carp from establishing a population in Lake Michigan are collaborative, involving numerous federal, state, and local agencies. This effort requires the synchronization of structural, chemical, biological, and other methods to be effective. In order to achieve success, all federal, state and local entities must apply their authorities, capabilities and resources as part of a comprehensive plan to address this challenge. The Corps will continue to apply its authorities and capabilities to achieve success.

Madam Chair, this concludes my testimony. I would be happy to answer any questions you or other members of the Subcommittee may have.

In: Troubled Waters: Asian Carp and the... ISBN: 978-1-61761-766-9
Editors: Deborah M. Maeda

Chapter 5

TESTIMONY OF MARC MILLER, DIRECTOR, ILLINOIS DEPARTMENT OF NATURAL RESOURCES, BEFORE THE SUBCOMMITTEE ON WATER RESOURCES AND ENVIRONMENT, HEARING ON "ASIAN CARP AND THE GREAT LAKES"

Thank you Madam Chair and members of the subcommittee, for this opportunity to testify on the actions Illinois Department of Natural Resources has undertaken since the early 1990's. I also will outline our action plans for the immediate future in our shared battle to prevent the spread of Asian carp to the Great Lakes.

Our commitment to this task has been and remains unwavering. We have been working closely with our partner states including Michigan and Wisconsin, and the federal agencies to develop effective control strategies. Illinois has contributed significant resources to controlling Asian carp over time. One example is that we were the local sponsor for the study, and testing of electric barrier system. Illinois contributed $1.8 million to this effort.

Most recently Illinois DNR served as the lead agency for the successful Rapid Response effort last December to prevent the migration of Asian carp when the electric barrier system was shut down for maintenance. The unified response of the Great Lakes States and Provinces was a shining leadership moment for our region, and a prime example of how a small group of committed people can make a difference.

This unprecedented effort demonstrated that Federal, Provincial, State, and Local partners can work together to ensure that this invasive species would not enter the Great Lakes and threaten one of the world's great ecosystems. Over 400 people worked together with contributions of supplies, equipment and crews from every member of the Basin. The Rapid Response team safely applied Rotenone to a six mile stretch of the Chicago Sanitary and Ship Canal. The USACE performed critical maintenance on the electric barrier system, and conducted cleanup and removal of 18,000 fish including one Big Head carp.

It is important to note that as we consider additional operations, the cost of this single action was $3,000,000 and would not have been possible without the substantial donations from the states and provinces and financial support of our federal partners. Thank you.

There are several lessons that we learned from this experience that I would like to share with the committee: first, meeting this challenge will require greater collaboration and levels of partnership. We must enlist the scientific and communication resources as well as the political leadership of every state and province in the basin to join in this effort.

Second, early outreach to key stakeholders, proactive communication strategies and operational transparency must continue to be maintained as we move forward with our framework strategy and operations.

Finally the collaborative approach that has been developed with our local, state, and federal partners is working very well and we believe represents the best model for future efforts.

I now wish to now outline the actions to control Asian carp that IDNR has identified to begin immediately or as soon as funding can be secured.

- We will conduct a targeted Asian carp removal operation throughout the entire Chicago Area Waterways System. This includes identification, containment and removal using conventional methods netting, electro fishing, commercial fishing, rotenone, etc.
- IDNR will contract with Commercial Fisherman to operate below the barrier system to reduce populations and propagule pressure.
- Intensive E-DNA monitoring, sampling and removal in hotspots of the Cal Sag. This includes the entire length of the Cal Sag below O'Brien Lock & Dam to the electric barrier.
- Participate with USACE efforts to refine the E-DNA technology to understand population densities and other factors.

- In the next 90 days IDNR will conduct a survey of all retail live bait locations to determine that live Asian carp minnows are not being sold in NE Illinois.

We have also identified several longer term actions that we are proposing as well:

- Prepare for Rapid Response contingency operations, including training, advance procurement of supplies and necessary equipment.
- Co-Chair the Asian Carp Management and Control Implementation Task Force with USFWS. This plan outlines 133 different actions that will be deployed nationally in all watersheds where Asian Carp are a problem.
- Conduct tagged fish research into barrier effectiveness using Didson side scan sonar
- Enhance commercial markets for Asian carp and investigate requirements for use of Asian carp products for humanitarian relief purposes.

This is a problem that is not going to be solved by one state, or one agency. As a region we have a long and established history of using a proactive and collaborative approach. Our Great Lakes Region is stronger when we work together in partnership to solve common problems, and Asian carp is a national problem.

When we are divided, solutions to our problems can remain elusive. Illinois DNR looks forward to working with the other Great Lakes States and Federal Agencies in developing sustainable solutions to our common problem.

In: Troubled Waters: Asian Carp and the... ISBN: 978-1-61761-766-9
Editors: Deborah M. Maeda

Chapter 6

TESTIMONY OF REBECCA HUMPHRIES, DIRECTOR, MICHIGAN DEPARTMENT OF NATURAL RESOURCES AND ENVIRONMENT, BEFORE THE SUBCOMMITTEE ON WATER RESOURCES AND ENVIRONMENT, HEARING ON "ASIAN CARP AND THE GREAT LAKES"

Madame Chair and Members of the Subcommittee:

My name is Rebecca Humphries, and I am the Director of the Michigan Department of Natural Resources and Environment. I appreciate the opportunity to testify today about the looming catastrophe that we face if Asian carp become established in the Great Lakes. I also appreciate the members of the Michigan Congressional delegation joining us today and for their past work on this and other Great Lakes issues.

I have been a conservation professional for over 30 years, and my role within the Michigan Department of Natural Resources and Environment is to protect our resources while maximizing recreational opportunities. Allowing Asian carp to populate our Great Lakes will destroy the resource as well as recreational opportunities. We must act swiftly, collaboratively, and wisely to address this crisis.

Invasive species have already created havoc in the Great Lakes. Reports indicate that the cost of biological pollution from invasive species is both massive and rising. In the Great Lakes, total costs for treatment and control of zebra mussels alone reach $100 million per year. The Great Lakes Fishery

Commission reports that for sea lamprey, the program requirements are on the order of $30 million per year.

Invasive species have profoundly changed the ecosystem of the Great Lakes, significantly impacted the Great Lakes sport and commercial fisheries and have hampered recreation, all of which have a negative effect on Michigan's economy.

Let me give you an example. Lake Huron once had a vibrant salmon sport fishery, with hundreds of charter boats attracting thousands of anglers each year to ports up and down its long coastline. Fishing derbies attracted additional anglers who launched their boats or kept their boats at local marinas. But invasive zebra and quagga mussels (Eurasian invaders) have caused the collapse of the salmon population, and thus the sport fishery. Gone are the fishing derbies, charter boaters have left the ports, and anglers have moved elsewhere. This was a several hundred million dollar industry, and it is gone.

Michigan has taken aggressive steps to stop the further spread of these foreign invaders, including:

- Requiring that Great Lakes ships adhere to ballast water management practices established by the shipping industry,
- Enacting legislation requiring all ocean-going ships to obtain a permit for ballast water discharges. The permit specifies the use of an approved treatment system to prevent release of invasive species via ballast water,
- Taking legal action to address ballast water issues, including successfully defending our state laws in federal court and challenging federal agencies for their failure to appropriately use existing regulatory authority to act, and
- Administering state regulatory programs to control aquatic nuisance species in our lakes and rivers. These programs include restrictions on transport of invasive species of fish, establishment of a list of invasive species prohibited in Michigan, and participation in actions to control sea lamprey in Great Lakes tributaries.

Despite our best efforts, Asian carp are now at our doorstep.

Michigan has its own steps to prevent Asian carp from entering the Great Lakes. We contributed financially to construction of the electrical barrier in the Chicago Sanitary and Ship Canal. We prohibited possession of live Asian carp in the state, and we participated in the response actions in December 2009 that treated the Canal to remove Asian carp prior to maintenance of the electrical barrier.

I cannot stress the following in simpler terms: Once an invasive species gets established in the lakes, we cannot eradicate it, fully control its spread, or the damage it causes.

But the story of Asian carp does not need to be a legacy of destruction for our children. The Great Lakes community, including Governors, congressional delegations, local government officials, and citizens has proven that they can work together on difficult challenges. Yes, this is a formidable challenge, but together we can and must solve it.

The threat of Asian carp must be treated as a crisis and steps must be implemented immediately to address it. As early as 2003, scientists, governmental officials, and stakeholders were calling for ecological separation of the Great Lakes and the Mississippi River watershed, but we did not act quickly enough. Short-term fixes have become long-term projects. For example, the installation of the second electrical barrier took over six years, and it is still not fully operational. It took several years to ban the importation of Black Carp and Silver Carp under the Lacey Act. Bighead Carp are still not covered under that Act.

Because of our history of insufficient action, I submit that any long-term actions must only be to install redundant prevention measures.

I started by saying that we must act swiftly, collaboratively, and wisely to address the threat posed by Asian carp.

Here are my recommendations to meet those objectives:

We must immediately take all available measures, consistent with protection of public health and safety, to prevent the migration of Bighead and Silver Carp into Lake Michigan, including:

- Closing and ceasing operation of the O'Brien Lock and the Chicago Lock until a permanent ecological barrier is constructed between the Great Lakes and the Mississippi River watershed. The Army Corp of Engineers must have the authority to close the locks on emergency and permanent bases if necessary,
- Initiating studies to be completed by the end of this year to examine the feasibility of transferring cargo via other transportation systems,
- Operating other water control structures near Lake Michigan — at the O'Brien Lock, the Chicago Controlling Works, and the Wilmette Pumping Station in a manner that will not allow fish to pass into the Lake,

- Installing interim barriers at other locations this year, including barriers between the Des Plaines River and the Canal and in Indiana Harbor and Burns Ditch from the Grand Calumet and Little Calumet Rivers to eliminate the potential for flooding between the two watersheds,
- Completing additional studies related to the biology/ecology of the carp and predictive models to determine the areas at highest risk for colonization in the Great Lakes, including estuaries and bays, drowned river mouths, and river systems,
- Providing additional dollars for continuous monitoring of carp based on risk analyses, with funding on reserve for chemical treatment used as a rapid response mechanism as warranted,
- Communicating with the states on actions and data in a timely manner,
- Operating electrical barrier 2a at optimum voltage and completing electrical barrier 2b this year, and
- Developing and implementing plans for a permanent solution to the problems that would ecologically and physically separate the carp-infested waters of the Mississippi River watershed from the Great Lakes.

We must also develop a proactive campaign to educate the public about the risks and dangers of Asian carp so that they do not get hurt or unknowingly (or knowingly) spread these dangerous fish into inland waters.

We all treasure the Great Lakes and share a commitment to their continued vitality. We must now all share a similar commitment to move aggressively forward to stop the spread of Asian carp. The Great Lakes states may have challenging discussions on specific actions, but that should not stop us from moving forward. Allowing Asian carp to populate our Great Lakes will destroy the resource and the recreational opportunities they provide us.

President Obama and Congress have given the restoration and protection of the Great Lakes new hope with the infusion of $475 million through the Great Lakes Restoration Initiative. This work will all be in vain if Asian carp are allowed into the Great Lakes.

My submitted testimony has additional attachments.

Thank you, and I would be happy to take questions from the Committee.

January 19, 2010

The President
The White House
Washington, D.C. 20500

Dear Mr. President:

We write to request an immediate summit between Great Lakes governors and senior White House officials to identify a rapid response to the threat posed by Asian carp to the Great Lakes Given the recent discovery of carp DNA samples above the Chicago Sanitary and Ship Canal barrier and today's Supreme Court decision, it is essential that we quickly implement emergenc measures to protect the Great Lakes from carp and other invasive species.

Invasive species have already had a significant impact on the economies of our states, and the introduction of Carp into this already fragile ecosystem will be devastating. The Great Lakes are vitally important not only to the 35 million people that live within their basins, but to the world. The Great Lakes play a part in all aspects of the economy from manufacturing and transportation to trade and tourism, and thus the protection of these resources is critical to our state economies. Your pledge of $5 billion to restore and protect the Great Lakes and the recen $475 million down payment on that promise through the Great Lakes Restoration Initiative is welcome and critical, and we pledge to work with you to implement these restoration measures in a timely and effective manner.

We look forward to working with the White House to identify and implement a solution to this critical issue. Thank you for your consideration and we look forward to your response.

Sincerely,

Governor Jennifer M. Granholm
Michigan

Governor Jim Doyle
Wisconsin

STATE OF MICHIGAN
JOHN D. CHERRY, JR.
LT. GOVERNOR

November 13, 2009
Ms. Jo-Ellen Darcy
Assistant Secretary for Civil Works
108 Army Pentagon Room 3E446
Washington, DC 20310-0108
Dear Assistant Secretary Darcy:

I write to you today on a serious matter that is critical to the health of the Great Lakes. Recent information released by the U.S. Army Corps of Engineers indicates that Asian Carp are very close to the partially operational and incomplete electrical barrier in the Chicago Sanitary and Shipping Canal. To avoid the devastating effects of these invasive species in reaching the Great Lakes, specific investments and actions are essential to ensure that the barrier and other infrastructure are controlling the water pathways used by aquatic invasive species. Below I have listed specific actions that need to be continued or initiated.

First, electrical barrier 2B needs to be completed and operational as soon as possible. The design for the electrical barrier in the canal consists of two parts, 2A and 2B. Of the two parts, to date only barrier 2A is operational, leading to serious problems with turning it off for maintenance. Barrier 2B is needed both for full protection from carp dispersal and to provide for maintenance of the electrical barrier as reqnired. As an interim measure, Michigan has supported an emergency response plan to protect the Great Lakes dwing a period when the existing barrier may be deactivated for maintenance, however, these emergency response efforts falls short of the long-term solution needed. Michigan was pleased to see funding to continue operation of portions of the electrical barrier in the canal and to continue construction of permanent barriers in the fiscal year 2010 Energy and Water Appropriations Act. This work must proceed expeditiously.

Second, safety studies must be quickly completed and procedures put in place to allow for full power operation of barrier 2A. To date, the barrier is operating at only half of its potential due to incomplete safety studies on its operation. Present operations may be enough power to deter adult Asian Carp from passing through the barrier but may not be enough to deter small fish

Third, a physical barrier, even if temporary, needs to be constructed as soon as possible between the Des Plaines River and the canal to eliminate the potential for flooding between the two waterways. Flooding could allow movement of fish, including Asian Carp, into the canal and vice versa. If the one-year period of authority for this action provided in the Energy and Water

Appropriations Act is insufficient, or if additional appropriations are needed to implement timely actions, I urge you to advise our Congressional delegation immediately so that they can continue to support protection of the Great Lakes from Asian Carp. I note it is important to the overall effort in the area of the canal that funding for physically separating the Des Plaines River and the canal not be re-programmed from the critical work on the electrical barrier itself.

Invasive species, such as Zebra Mussels, Round Goby and Eurasian Ruffe, have already demonstrated devastating environmental and economic effects on both the Great Lakes and Mississippi River watersheds. Further work on a permanent solution to prevent the dispersal of aquatic invasive species needs to proceed in a timely manner. Michigan was pleased to see funding to continue a study, including consultations with key partners on the range of options and technologies to prevent the inter-basin transfer of aquatic invasive species in the fiscal year 2010 Energy and Water Appropriations bill. This is important work for protection much of the U.S. from invasive species dispersal and I would urge you to make the review of issues related to the Chicago Sanitary and Shipping Canal a top priority.

Thank you for your continued attention to this matter. Please let me know if there is anything more the state of Michigan can do to help with this critical work.

Respectfully,

John D. Cherry, Jr.
Lieutenant Governor

STATE OF MICHIGAN
OFFICE OF THE GREAT LAKES
LANSING

JENNIFER M. GRANHOLM
GOVERNOR

KEN DEBEAUSSAERT
DIRECTOR

November 25, 2009

To Emergency Response Group:

The recent eDNA evidence indicating that Asian Carp may have made their way past the electrical barrier in the Chicago Sanitary and Shipping Canal raises an even higher level of threat to the ecological integrity of the Great Lakes and puts a $4.5 billion sport and commercial fishery at risk.

While the state of Michigan has supported the Emergency Response Plan developed earlier, in light of the new evidence, we strongly urge you to consider taking additional preventive measures including:

- Emergency closing of the locks in the Cal-Sag Channel, chemical applications in this area, and monitoring until the eDNA evidence is confirmed or ruled out
- Other alternative testing to confirm the eDNA evidence
- Expanding the chemical applications to additional areas where eDNA evidence indicates the presence of Asian Carp

Since the emergency response is planned to begin December 2, 2009, impacts on commerce from emergency closure of the locks is minimal compared to the potential economic harm that could be caused by Asian Carp's introduction. If the emergency response confirms the eDNA evidence, additional and more permanent actions will be necessary to provide protection for the Great Lakes.

Actions called for in the Great Lakes Regional Collaboration Strategy related to options for permanent separation of the Great Lakes and Mississippi River system must be expedited. Recently, Michigan's Lt. Governor John D. Cherry wrote the Undersecretary of the Army to urge additional actions to protect the Great Lakes, including the completion of the second portion of the new barrier, full utilization of the existing barrier, now operating at minimal levels, creating a physical barrier to block Asian Carp from entering via other waterways during times of flood, and using all existing authorities to explore all options to block the Asian Carp.

As you know, the state of Michigan has long-supported efforts to block Asian Carp from entering the Great Lakes, including prohibiting possession of live Asian Carp here. We have provided direct financial support for the electrical barrier in 2004, have continuously supported federal authorizations and appropriations, and as you know; we are providing manpower, materials, and equipment for the upcoming emergency response along with other states and provinces.

But ultimately at this critical moment, the awesome responsibility to protect the Great Lakes from the introduction of Asian Carp is yours. The actions you take must err on the side of protection of our Great Lakes.

Sincerely,

Ken DeBeaussaert
Director
517-335-4056

cc: **Mr. Cameron Davis, U.S. Environmental Protection Agency**
Ms. Donna Stine, Governor's Office
Mr. Steven E. Chester, Director, MDEQ

Ministry of Natural Resources

Office of the Minister

Room 6630, Whitney Block
99 Wellesley Street West
Toronto ON M7A 1W3
Tel: 416-314-2301
Fax: 416-314-2216

Ministère des Richesses naturelles

Bureau du ministre

Édifice Whitney, bureau 6630
99, rue Wellesley Ouest
Toronto (Ontario) M7A 1W3
Tél.: 416-314-2301
Téléc.: 416-314-2216

Ontario

MNR4046MC-2010-91

February 2, 2010

State of Michigan
Office of the Governor
PO Box 30013
Lansing, Michigan 48909
USA

Attention:
Governor Jennifer M. Granholm

Dear Governor Granholm:

I am writing to express Ontario's concerns regarding the potential spread of Asian Carp into the Great Lakes Basin. I understand that the Director of Michigan's Department of Natural Resources and Environment, Rebecca Humphries, will be testifying before the subcommittee of Water Resources and Environment of The House of Representatives on February 9, 2010.

We have been advised that this hearing presents Ontario with an opportunity to ensure that our concerns regarding the negative environmental and socio-economic impact of Asian carp entering the Great Lakes are considered by the Committee. The people of Ontario appreciate your office's assistance in submitting Ontario's letter as part of Michigan's testimony. Canada and the United States have a long history of shared stewardship of the Great Lakes and we wish to continue to build on this already strong relationship. Our concerns are described below.

The potential introduction of Asian carp into the Great Lakes poses an imminent and substantial threat to the environment and economy of Ontario. As such, it is vital to the Province that immediate action be taken to prevent the introduction of this invasive species. Preventing the introduction of Asian carp into the Great Lakes, is far preferable to, and less costly than, attempts at eradication or control. Prevention measures must be the first line of defence before eradication, which is often only partially effective following establishment of an invasive species.

There will be profound consequences for Ontario if this invasive species is introduced to the Great Lakes. The waters of the Great Lakes are so interconnected that irreparable harm to U.S. states will cause similar harm to Ontario. What happens in Illinois' waters can and will affect the sustainability of the Great Lakes' fish populations shared with Ontario. Preventing the introduction of Asian carp into the Great Lakes and protecting our shared fishery resources is therefore of great national and international importance.

As shown in the map below, approximately 40% of the shoreline of the Great Lakes and 36% of the waters of the Great Lakes lie within the boundaries of Ontario

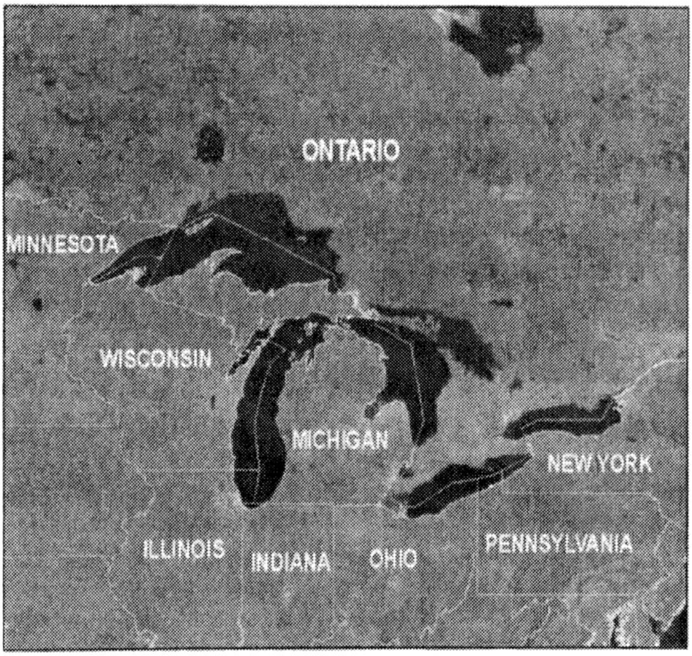

The fishery resources of Ontario's Great Lakes waters have major social, environmental and economic importance to the Province, and all persons who live in the Great Lakes Basin. The economic value of the sport and commercial fisheries in Ontario's portion of the Great Lakes is substantial. Based on 2005 statistics, the combined value of the recreational and commercial Great Lakes fisheries is approximately $643 million (Canadian). These fisheries and the enormous commercial and governmental investments in these fisheries are at severe risk, should Asian carp become colonized in the Great Lakes system. Lake Erie is likely to be the Great Lake most severely affected.

The sound management of the Great Lakes aquatic resources and water dependent natural resources is vital to the Province of Ontario. Over time. the Great Lakes fishery resources have been diminished and significantly altered through exploitation, degradation of habitat and the introduction or invasion of plant and animal life.

Cooperative decision-making is the best way to manage and conserve a fragile resource. As part of its efforts to manage its fishery resources, the governments of Canada and the Province of Ontario have recognized that coordination and cooperation with the United States is a critical element to ensure that fisheries are maintained in a sustainable way. It is for such reasons that state-provincial agreements, as well as bilateral treaties, have been entered into on water quality, water levels, water removals and fisheries, to ensure their coordinated and effective management and protection.

To date, the Great Lakes jurisdictions have recognized a responsibility to take action within their borders. Ontario and the Great Lakes states have acted together to address the potential threat of the introduction of live Asian carp into the Great Lakes Basin. In 2006 and 2004, respectively, Canada and Ontario amended their regulations to make it illegal to possess, buy or sell live Asian carp within the Province, and we have enforced these regulations. However, this is not enough and does not address the threat that Asian carp pose as they travel up the Mississippi River. Jurisdictions must act quickly and in a cooperative manner as has been done in other contexts.

Previous experiences have shown that the introduction of an invasive species can have devastating affects. For example, zebra mussels, another invasive species, were introduced into the Great Lakes in the mid-1980's. Zebra mussels significantly changed the nature of the Great Lakes ecosystem, increasing water clarity, disrupting the food web and affecting fish habitat by altering the structure and composition of critical spawning habitat. Overall, the impact was a reduction in the amount of food available to native species. In fact, zebra mussels have caused drastic declines in the native Great Lakes clams. An estimate of the amount that has been spent to address problems related to zebra mussels in both the United States and Canada in the Great Lakes Basin is between \$3 and 7.5 billion (Canadian). Between 1989 and 2004, Ontarians alone spent \$120 million (Canadian) on zebra mussel control.

The sea lamprey has had equally devastating affects on the once commercially and environmentally significant Lake trout and other fish. Decimation of Lake trout began when sea lamprey entered Lake Erie and the upper Great Lakes in the early 1920's. So significant was the negative impact that the U.S. and Canadian federal governments entered into the 1954 Convention on Great Lakes Fisheries, one of the purposes of which was to develop a united attempt by the affected U.S. and Canadian jurisdictions to battle the sea lamprey.

The cost of sea lamprey control measures has been substantial; approximately \$20 million (U.S.) annually, of which Canada's contribution in 2009 was \$8.1 million (Canadian). This annual and continual requirement to spend money in an attempt to control sea lamprey has not resulted in their eradication.

It is vital that immediate action be taken to prevent the introduction of Asian carp into the Great Lakes system in order to prevent serious environmental and economic damage to the Province of Ontario.

The Great Lakes are a treasure shared by both Canada and the United States. I appreciate the assistance that your staff have provided to ensure that Ontario's interests and concerns are considered. By working together we can ensure our actions support the concerns expressed by Michigan and others and also ensure that the bi-national aspect of the issue is before the subcommittee.

Sincerely,

Linda Jeffrey
Minister of Natural Resources

c: Premier Dalton McGuinty
Rebecca Humphries, Director, Michigan Department of Natural Resources and Environment

In: Troubled Waters: Asian Carp and the... ISBN: 978-1-61761-766-9
Editors: Deborah M. Maeda

Chapter 7

TESTIMONY OF MATT FRANK, SECRETARY OF THE WISCONSIN DEPARTMENT OF NATURAL RESOURCES, BEFORE THE SUBCOMMITTEE ON WATER RESOURCES AND ENVIRONMENT, HEARING ON "ASIAN CARP AND THE GREAT LAKES"

INTRODUCTION

Chairwoman Johnson and members of the Subcommittee, thank you for the opportunity to appear before you today to discuss our shared efforts aimed at protecting the Great Lakes from aquatic invasive species – in this case Asian Carp. My name is Matt Frank and I am the Secretary of the Wisconsin Department of Natural Resources. I am pleased to submit this testimony on behalf of the WDNR and my boss, Wisconsin Governor Jim Doyle. I want to start by thanking Representative Oberstar for his leadership on many Great Lakes issues as Chair of the House Transportation and Infrastructure Committee. I also want to applaud Representative Kagen for his leadership as a member of this Subcommittee and I would also like to recognize Representative Petri who serves on the full Committee on Transportation & Infrastructure.

The Great Lakes in Perspective

The Great Lakes are a treasure of international significance. They contain approximately 20% of the world's surface freshwater, and 95% of North America's. One in three Canadians and one in 10 U.S. residents depend on the Great Lakes for their water. More than *35 million U.S. residents* and 8 million Canadians live, work, and recreate in, on or by the waters of the Great Lakes Basin.

The Great Lakes regional economy and, indeed, our nation's depend on the Great Lakes. For example, the Great Lakes provide water for 70 percent of U.S. steel production. The Lakes provide transportation for almost 200 million tons of international and interlake cargo—indeed, the lake carriers can tell you how much transport tonnage they lose for each inch of water lost. One-third of all the boats registered in the United States are in the Great Lakes States and boating alone supports over 250,000 jobs. Overall, our region generates nearly 30% of our nation's gross domestic product and about 60% of all U.S. manufacturing. Water is also used for hydro-power on both sides of the border. All of these different uses depend on the lakes in different ways as a source for clean, abundant fresh water.

Wisconsin boasts a $13 billion tourism industry, much of those dollars are generated thanks to abundant, healthy water resources and in turn a popular recreational fishery.

Wisconsin waters of Lake Michigan and Lake Superior support a popular and thriving sport fishery which includes private anglers, licensed guides and charter captains. There are approximately 235,000 anglers who fish 3.7 million days each year primarily for rainbow, brown and lake trout, chinook and coho salmon, walleye, smallmouth bass, strain muskellunge, and yellow perch. Wisconsin licensed 359 charter captains in 2009. Based on Wisconsin Department of Natural Resources (WDNR) creel surveys of major fisheries in 2008 (not all fisheries are covered), anglers harvested a minimum of 640,000 fish in Lake Michigan and 36,000 fish in Lake Superior.

Sport fishing in Wisconsin waters of Lake Michigan and Lake Superior generated $419 million in economic activity and supported 5,000 jobs in Wisconsin alone, based on a comprehensive survey conducted in 2006 by the United States Fish and Wildlife Service and the Department of the Census and an economic analysis done by the American Sportfishing Association.

Sustainable management and use of the Great Lakes can foster economic growth while protecting our environment. Conversely, we place our water resources, our environment and our economy at risk if we do not manage the

Lakes sustainably and do not keep our lakes at healthy levels. Therefore, we must be forward-looking to put in place effective policies that address today's issues and anticipate tomorrow's challenges. These policies should include immediately and effectively tackling our most pressing problems; putting in place an effective long-term water management framework; and, developing a robust research and information-sharing regime to encourage adaptive management.

Restoring and protecting the Great Lakes is a persistent challenge requiring myriad and collaborative actions across all levels of government. Required actions are not easily isolated from one another, nor should they be, and we must look at all of the challenges that face our Great Lakes.

GREAT LAKES REGIONAL COLLABORATION STRATEGY

Highlights the Challenge of Aquatic Invasive Species

When I testified at your field hearing in Green Bay in the spring of 2008, I noted with pride the collective achievement of Great Lakes' region in producing the Great Lakes Regional Collaboration (GLRC) Strategy to achieve our shared goals. The framework for the GLRC was based on the nine priorities that the Great Lakes Governors outlined in 2003. The process started with a Presidential Executive Order and included our regional leaders—Governors, Mayors, Members of Congress and Tribal leaders—as well as non-governmental groups and hundreds of committed citizens. The process united us as never before toward our shared goals of protecting and restoring our nation's water belt--the Great Lakes. And, it provided a shared vision of near-term steps that could put us on a path toward a restored water belt--a healthy water belt to power our nation's economy and support a robust environment.

That promise is now being brought to reality thanks to the hard work of those same Great Lakes stakeholders and the welcome federal commitment of significant resources to support the strategy through the Great Lakes Restoration Initiative. As you know, this Initiative was funded by this Congress at the full $475 million requested by President Obama for Federal Fiscal Year 2010.

Aquatic Invasive Species Control Key Strategy

While we are extremely grateful to Congress and to the Administration for your support of the GLRI, it is noteworthy that over $60 million of the $475 million for the Initiative is being expended just to combat aquatic invasive species. Indeed, aquatic invasive species (AIS) continue to pose one of the most serious threats to the Great Lakes ecosystem. An average of one new species is discovered in the Great Lakes ecosystem every eight months, and once present, eradication is often impossible. Prevention is vital to stemming ecosystem impacts from new invasive species. And, because AIS easily transfer from watershed to watershed, it is absolutely critical that comprehensive national action be taken to combat the spread of AIS.

Chicago Ship Canal Just One Vector for Invasives

Yesterday's White House Asian Carp Summit was a very important step in our collective efforts to keep Asian Carp out of the Great Lakes. Your support of additional federal revenue for this task would also be very appreciated. For several weeks now, we have been raising several issues where swift action is needed. Some of the key issues where we continue to focus our attention and urge concrete action include:

Achieving a total "ecological separation" of the Mississippi-Illinois River System from the Great Lakes drainage.

Background: Like Michigan, Wisconsin is gravely concerned about Asian Carp and other AIS movement into Lake Michigan, but unlike Michigan we also have to live with AIS movement downstream into the Mississippi River system. The current electrical barrier provides no protection against downstream movement, so "ecological separation" is the only effective option for Wisconsin. In Chicago, this means infrastructure changes in the **Chicago Waterway System** (CWS) such that there are no direct hydrologic connections between the Illinois Sanitary and Ship Canal and Lake Michigan.

To ensure that Asian Carp do not become established in Lake Michigan before "ecological separation" is completed, we also want swift action in these areas:

Operate the existing Electrical Dispersal Barrier system at maximum effective power and expedite completion of the proposed Barrier IIB.

Background Congress first directed the COE to deal with the problem of invasive species movement in the CWS in 1996 and they have slowly been developing a system of 3 electrical barriers at a bottleneck location on the Chicago Sanitary and Ship Canal (see attached diagram). The first was a low power "demonstration" barrier (Barrier I) which did not begin operation until 2002. After lengthy wrangling over funding, a second, more powerful dual barrier has been incrementally constructed. Testing began on the first barrier of the new dual system (Barrier IIA) in March 2006 and finally began full time operation in April 2009. Due to safety concerns it is still not being operated at a voltage that is sufficient to repel all sizes of Asian Carp. COE has stated that it has received funding for the second barrier of the new dual system (Barrier IIB) and that construction is underway and will be completed by September, 2010. COE states that operational and safety testing must be completed before Barrier IIB can start operation and provides no timetable for completion of the tests. In the years since Congress first instructed the COE to deal with this issue, progress has been slow. In 2004, despite the fact that this interstate waterway is the responsibility of the Federal government, the eight Great Lakes states contributed $575,000 needed to fill a gap in funding to construct the barrier (Wisconsin contributed $68,000 from WNDR).

Comprehensively monitor the Chicago Sanitary and Ship Canal and all connected waterways for the presence and location of bighead and silver carp using the best available methods and techniques.

Background The urgency of creating "ecological separation" and the severity of interim measures depends to some extent on the level of Asian Carp infestation in the CWS above the electric barrier. So it makes sense to routinely conduct comprehensive fish sampling to monitor the level of infestation. Over the years, Illinois DNR, US Fish and Wildlife Service, and several other agencies have done varying levels of Asian Carp monitoring primarily downstream, but also occasionally at or above the electric barrier. Monitoring is usually done with normal fish sampling gear such as electrofishing boats, nets or radio tracking. Starting in August, 2009, COE began looking for Asian Carp DNA in water samples using an experimental technique they call "environmental DNA" (eDNA) testing. This new testing has not been calibrated – *it is not known what level of Asian Carp infestation will trigger a positive eDNA test* - but this testing has found evidence of Asian Carp DNA at several locations above the barrier. Illinois DNR has recently been trying to organize a coalition of state and federal agencies – including Wisconsin DNR - to contribute resources for an actual fish sampling monitoring program for 2010.

Eradicate any bighead or silvercarp discovered in these waters.

Background Again, to ensure that Asian Carp do not become established above the electric barrier and escape into Lake Michigan before "ecological separation" can be created, it will be necessary to eradicate any populations of Asian Carp discovered by the monitoring program described in #3a. Currently the only feasible way to control Asian Carp is to chemically poison all the fish in the infested location. The chemical used is called rotenone and while deadly for fish is not any threat to human health or other land or water animals. These chemical treatments however, can be logistically difficult and expensive. For example, in December 2009, Illinois DNR sponsored a chemical treatment of a 5 mile stretch of the **Chicago Sanitary and Ship Canal (CSSC)** just below the electric barrier to keep Asian Carp from passing the barrier while it was down for maintenance. Illinois DNR sought financial and staff assistance from other agencies and the final effort included 250 staff from many federal and local agencies, states around Lake Michigan and even Ontario and Quebec, at a total cost of probably $5 million. It is likely that treatments in other areas of the CWS should Asian Carp populations be discovered would be of comparable magnitude – however in the future it is unclear of any of the other participating agencies would be able to repeatedly donate resources to frequent treatments.

Expedite completion of the barriers between the CSSC and the Des Plaines River far enough upstream to ensure that Asian Carp cannot cross over during flood events.

Background The location of the electrical barrier on the CSSC is above the confluence of the Des Plaines River which means that Asian Carp coming upstream from the Illinois River would have unimpeded access to the Des Plaines River. Unfortunately, the Des Plaines River flows right next to the CSSC for a distance of approximately 13 miles separated only by a narrow strip of land. During flooding events (such as occurred in summer, 2008), water – and potentially Asian Carp - can move between the Des Plaines and the CSSC above the barrier which would give them direct access to Lake Michigan. Nothing was done about this potential barrier breach until December, 2009 when EPA announced it would provide $13 million of its Great Lakes Restoration Initiative funding to allow construction of levees and fences to help separate the Des Plaines and CSSC. While a worthy effort, it is unclear if the funding provided or the final engineering plans will be sufficient to provide an adequate fish barrier.

Quickly assess the impacts of actions such as closing the locks and modifying sluice gate operations. Discuss and evaluate those impacts and agree to immediate modifications to lock and sluice gate operations that significantly minimize the chances of Asian Carp movement but also are protective of public health, safety and significant commerce issues.

Background In various court filings, COE, State of Illinois and the Chicago Metropolitan Water Reclamation District have sincerely argued that there are significant flooding and commercial barge traffic economic implications to preliminary injunction requests for immediate lock closures and changes in water diversion operations. At the same time, the states of Michigan, Wisconsin, Minnesota and Ohio have all sincerely argued that their fishery interests worth $7 billion are also at grave risk should Asian Carp become established in the Great Lakes. Assuming that these both represent extreme positions, it would seem the parties could have a fair discussion of the impacts, and see if there are any reasonable changes in lock or water diversion operations that could be taken on an interim basis that would reduce the risk of Asian Carp movement but still reasonably address flooding and barge movement issues.

Complete construction of passive Asian Carp barriers on the Mississippi River.

Background Asian Carp are extremely abundant in the Mississippi River just south of Wisconsin, and there are no impassable barriers to keep them from spreading up into Wisconsin and Minnesota. In 2004, Minnesota, Wisconsin and US Fish and Wildlife Service paid for a joint study of barrier options for the Mississippi River which made some specific recommendations for installation of passive barriers (*http://files.dnr.state.mn.us/natural_resources/invasives/aquaticanimals/asiancarp/umrstudy.pdf*). Generally passive barriers include lights, sound, bubble screens and chemical attractants installed at bottleneck points such as locks. The study also recommended these passive barriers be installed at two neighboring lock and dams, and that active control and removal programs (eg localized chemical treatments or commercial harvest) be initiated should Asian Carp penetrate the downstream barrier. At the time it was estimated such a barrier system would cost about $25 million. COE was asked to include this as part of their infrastructure redevelopment, however no action has been taken to date (it was not included among the COE ARRA projects for example).

Federal Actions Needed to Prevent Other Invasive Introductions

As Governor Doyle, myself and many other in the Great Lakes region have requested repeatedly, the federal government must move swiftly under its existing authorities to require improvement for ballast water management including practices for those ships declaring no ballast on board to forestall the introduction of new invasive species to the Great Lakes.

A system of state by state regulations will not be nearly as effective and is clearly less desirable than a consistent and clear federal solution. In Wisconsin, we have already acted to require a state ballast water permit for ships coming into our ports but this is not a substitute for meaningful federal action.

We have commented on the draft Coast Guard ballast water permit (see Attachment A) and have urged support for federal legislation in the past (see Attachment B).

Federal support is also needed to strengthen rapid response capabilities. Legislation is needed to prevent the introduction and spread of harmful species via the trade in live organisms and other vectors. Finally, Congress should provide full funding for the Great Lakes Fishery Commission's sea lamprey control program and other state and regional programs under the National Invasive Species Act.

CONCLUSION

More than 180 non-native species have invaded the Great Lakes, damaging water-dependent industries, threatening valuable fish and wildlife resources, and costing the region an estimated $5.7 billion annually.

The region must remain vigilant to prevent new invasive species from entering the Great Lakes and causing long-term, irreversible damage. The imminent threat of Asian carp devastating the region's $7 billion sport fishing industry underscores the urgent need for action to safeguard the Great Lakes against the threat from aquatic invasive species.

The continued health and availability of Great Lakes water in this region is critically important for our nation's environment and economy. Fortunately, we have a strong partnership and tools that we can use to ensure our future. Ms. Chairwoman and members of the Committee, our pledge to you is that we will continue to work with you to ensure that we make real progress on preventing Asian Carp and other aquatic invaders from entering and further compromising the health of our Great Lakes. This is our responsibility to our citizens, our children and our grandchildren.

Thank you, Ms. Chairwoman.

In: Troubled Waters: Asian Carp and the... ISBN: 978-1-61761-766-9
Editors: Deborah M. Maeda

Chapter 8

TESTIMONY OF DAVID M. LODGE, DIRECTOR, CENTER FOR AQUATIC CONSERVATION, BEFORE THE SUBCOMMITTEE ON WATER RESOURCES AND ENVIRONMENT, HEARING ON "ASIAN CARP AND THE GREAT LAKES"

Madam Chairwoman and Subcommittee members, I am honored to have the opportunity to participate in this hearing. I thank the subcommittee, especially Chairwoman Johnson, for the invitation to testify.

I am a biologist. I received a D.Phil. from the University of Oxford as a Rhodes Scholar. I come to the issue of Asian carp in the Great Lakes from the perspective of an active researcher on many kinds of invasive species in many places worldwide. I have been working on invasive species for 26 years, supported by private funding as well as funding from the National Science Foundation, the Environmental Protection Agency, the National Oceanograpic and Atmospheric Administration, and, of greatest relevance to today's hearing, the Army Corps of Engineers. I have co-edited two books and have authored or co-authored at least 150 published scientific papers. At the University of Notre Dame, I am the Director of the Center for Aquatic Conservation, Director of Notre Dame's new Environmental Change Initiative, and a professor of biology. My colleagues, collaborators, and I have many on-going research projects on various aspects of invasive species. The topics of particular relevance to this hearing include: (a) forecasting the spread and the environmental and economic impact of many aquatic nuisance species; (b) developing new species risk assessment (screening) protocols to allow private sector and agencies to prevent the importation and sale of harmful species while leaving unaffected the sale of benign species; (c) combining economic and ecological risk analyses to guide allocation of resources among management options; and (d) developing new genetically-based surveillance technologies for environmental protection.

I also bring considerable experience at the science-policy interface to my studies on Asian carps. I am a past Chairman of the national Invasive Species Advisory Committee. I was also the chairman of a committee appointed by the Ecological Society of America (the primary scientific society for biologists who study the environment), to write an assessment of the science and policy of invasive species (Lodge et al. 2006). The current state of science, economics, management, and policy on invasive species was assessed by many scientists and economists in a recently published book my colleagues and I edited (Keller et al. 2009). Thus, I can represent a consensus of views from the scientific and social sciences on invasive species biology and economics.

I was last before this subcommittee in March 2007 to testify about the impact of ship-borne invasions in the Great Lakes. In that testimony, I pointed out that while ships were the major contributor of alien species to the Great Lakes, other pathways were also important, including canals. Today, I am here to talk about the Chicago Sanitary and Ship Canal and the threat it poses as a conduit for the dispersal of potentially harmful species into the Great Lakes. The threats urgently at hand are silver carp and bighead carp, but the canal represents a potential highway to environmental havoc for many species that pose a high risk to both the Great Lakes and the Mississippi basins.

In particular, I will address the following topics in turn: what my collaborators and I have learned in the last few months about how close to Lake Michigan are the silver and bighead carps in the Chicago canal (the Chicago Sanitary and Ship Canal and connecting waterways); how many carp would it take to launch an invasion of Lake Michigan; how important the Chicago canal is relative to other pathways by which silver or bighead carps could gain access to Lake Michigan; the likelihood of substantial damages to Great Lakes fisheries, recreation, and economies if silver and/or bighead carp invade the Great Lakes; the importance of the Chicago canal as a conduit for many species dispersing in either direction; and what management actions I believe the available evidence suggests should be more seriously considered and/or pursued for the Chicago canal.

In the Chicago canal, how close are silver carp and bighead carp to Lake Michigan?

In this section, I describe recent genetically-based surveillance results on silver carp (*Hypophthalmichthys molitrix*) and bighead carp (*H. nobilis*) by a team of researchers, including Mr. Lindsay Chadderton, Dr. Andrew Mahon, and Dr. Christopher Jerde, and me. Mr. Chadderton is the Director for Aquatic Invasive Species, Great Lakes Project, The Nature Conservancy, has authored or co-authored at least 17 published scientific papers, and has over 15 years of experience in both New Zealand and the United States employing surveillance and rapid response efforts in the management of invasive species. Dr. Mahon is a Research Assistant Professor of Biological Sciences at the University of Notre Dame, has authored or co-authored at least 13 published scientific papers, and is an expert on ecological genetics. Dr. Jerde is a Research Assistant Professor of Biological Sciences at the University of Notre Dame, has authored or co-authored 14 published scientific papers, and is an expert on analyzing and interpreting ecological data, including presence-absence data of organisms like the data that we have been collecting on Asian carp. Because this section represents the collective work of the individuals listed above and additional laboratory and field technicians, I use "we" in this section to describe our results. Nevertheless the interpretations and views expressed in this section and subsequent sections are my own.

We recently provided a detailed description of our methods and results in a declaration to the Supreme Court (Lodge 2010). Here I provide a summary and results that are updated to 2 February 2010).

To overcome the well-known limitations of traditional fisheries sampling gear in situations where the target fish is rare (like the situation at an invasion front), we have used a non-traditional method. In early 2009, we adapted and improved the environmental DNA (eDNA) method of Ficetola et al. (2008) so that we could detect the eDNA of silver carp and bighead carp, collectively referred to as Asian carps, in the Chicago canal. By eDNA, we mean DNA in bits of tissue routinely left behind by all organisms. In the case of these fishes, the DNA could be contained in cells from multiple sources including slime, feces, and urine. Specifically, we identified and used species-specific molecular markers for silver carp and for bighead carp. These markers recognize sequences of DNA that occur only in silver carp or only in bighead carp and not in any other species of organism. Beginning in summer 2009, the Army Corps of Engineers began to financially support our use of the environmental DNA (eDNA) tool as potentially the best available technology to detect the presence of silver carp and bighead carp where they occur at low abundance. As of our 20th and most recent sampling trip on 8 December 2009, we have sampled a large portion of the Chicago canal (Figure 1). While we have sampled some reaches of the canal more than three times (to confirm detections of eDNA from silver and/or bighead carp), we have not yet sampled some northerly reaches of the waterway even once, nor have we sampled any Indiana portions of the Grand Calumet and Little Calumet rivers (Figure 1).

During 15-16 December 2009, our field and laboratory protocols were examined in detail by a four member team of experts. This Quality Assurance audit team was led by the Environmental Protection Agency, and was independent from the research team and the Army Corps of Engineers. An observer from the Army Corps of Engineers was also present. The review team scrutinized all the reports that we have provided to the Army Corps of Engineers and our laboratory notebooks, inspected our laboratory at Notre Dame, observed and discussed with us all the details of our eDNA protocols, and provided us with blind samples to process. In their Summary, the team confirmed that our genetic markers detect only the target fish species, endorsed our field and laboratory protocols, acknowledged that our methods minimized the possibility of reporting false positive results, and concluded the following: "Our team believes that the eDNA method you are using is sufficiently reliable and robust in reporting a pattern of detection that should be considered actionable in a management context. We have a high degree of confidence in the basic PCR method you are using for detecting Silver and Bighead carp environmental DNA"

Additional confidence that the presence of eDNA indicates the presence of live Asian carps is provided by at least two additional lines of strong evidence. First, in more southerly portions of the Chicago canal, where Asian carp are known to be abundant from traditional methods, a large proportion of our water samples contain eDNA of one or both species. Of the seven collection trips below the electric barrier, silver carp DNA was detected from six trips and bighead DNA from five trips. The only trip in which DNA was not collected occurred on 3 August 2009 where most of the samples were collected just below the electric barrier and in the main channel instead of the backwaters, where we discovered DNA accumulates.

Fig. 1: Sampling effort in the Chicago canal (June-December 2009) by the University of Notre Dame and The Nature Conservancy. Additional sampling conducted in the Des Plaines River and the I&M Canal is not indicated on the map but is discussed in text.

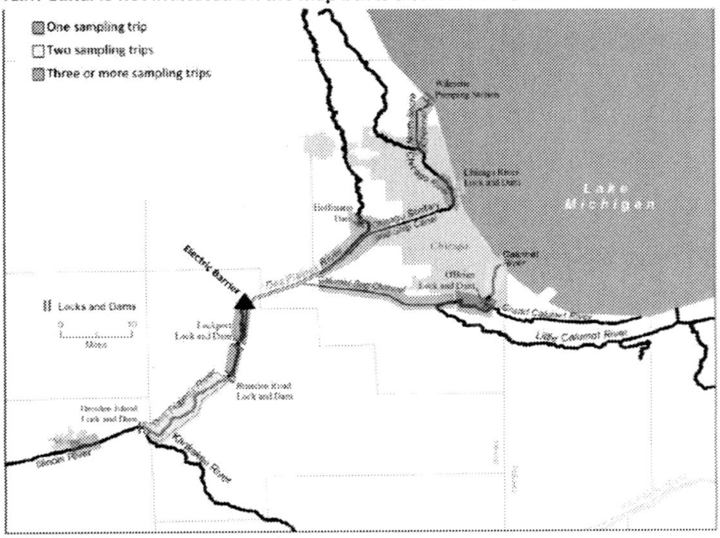

Fig. 2: Locations and sample dates of detection of eDNA for silver (S) and bighead (B) carps by the University of Notre Dame and The Nature Conservancy. Colors indicate results for both species combined. Des Plaines River and I&M Canal results are presented only in the text.

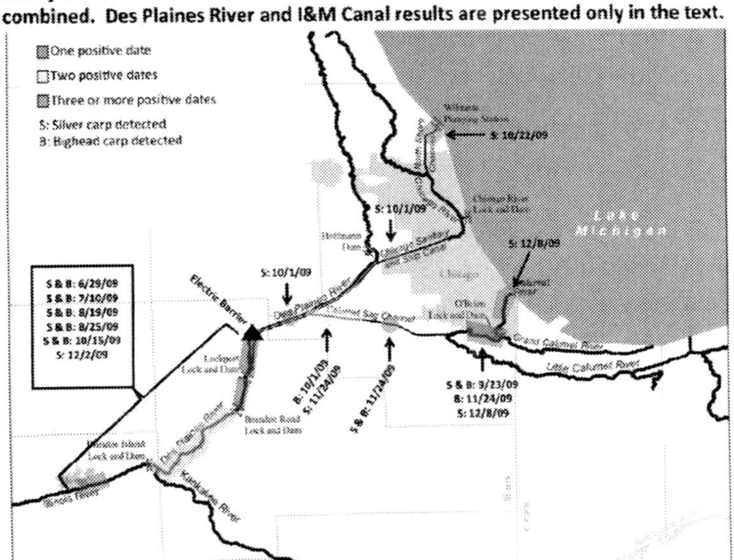

Second, when personnel from management agencies applied poison or traditional methods very intensively where we had detected eDNA, at two locations they saw or caught one silver carp and one bighead carp. At these locations prior to our eDNA studies, neither species was thought to occur. Given that sampling is being conducted at or near the invasion front (where the fishes are rare), it is not surprising that fishes have not been caught or seen at all the locations we have detected eDNA.

The most alarming of our eDNA evidence suggests that at least one silver carp has been in Lake Michigan near the mouth of the Calumet River, and at least one other silver carp has been within 100 hundred yards of the Wilmette Pumping Station (Figure 2). The closest to Lake Michigan that we have detected bighead carp is about 7.5 miles, about 1 mile south of O'Brien Lock & Dam (Figure 2). As Figure 2 indicates, we have detected additional silver carp and bighead carp in multiple locations at multiple times above the electric barrier.

The rate at which we have collected water samples, analyzed them, and reported results has met or exceeded the rates agreed under our cooperative agreement with the Army Corps of Engineers. We have collected a total of 577 two-liter water samples above the electric barrier. We have completed the processing of 84% and 82% of those samples for bighead carp and silver carp, respectively. Because the processing of water samples is much more time consuming than their collection, we continue to report results. It is important to understand, however, that the more recently reported results near Lake Michigan reflect only the order of our sampling and processing, not the movement of fish. Our sampling was not designed to detect any changes in the carps' distribution during the last few months of our sampling. Rather our results (Figure 2) should be understood as a snapshot of the carps' occurrence over the last few months.

Our protocols detect only whether some eDNA was present, not how much eDNA was present. Furthermore, even if we knew the initial concentration of eDNA in the water sample, we still would not be able to conclude anything about the absolute abundance of fish (e.g., number of fish per area) because we do not know how the abundance of eDNA relates quantitatively to the abundance of fish under even one given set of environmental conditions (e.g., temperature, current velocity). While these eDNA results cannot tell us how many individual fish of each species occurred of even exactly where or when, a positive result (i.e., the detection of eDNA) indicates that at least one fish has been present nearby our sampling location or just upstream (the water flows from Lake Michigan southward in all reaches of the Chicago canal) within the few hours before we sampled. We base this inference on laboratory experiments in which we were no longer able to detect fish eDNA in laboratory tanks 6 to 48 hours after the removal of a fish. We believe that under natural conditions in the waterway, the eDNA signal would degrade even more rapidly.

Without additional experiments, it is impossible even to conclude anything about the abundance of Asian carps in one location compared to another location with different flow conditions or other differences that might affect the presence or persistence of eDNA. For this reason, we do not compare results from the lower reaches of the canal to the upper reaches of the waterway where flow conditions are very different.

Confidence that eDNA indicates the presence of at least one individual live fish of a target species applies especially strongly to locations where we have detected eDNA on multiple sampling trips separated by multiple days (during which any eDNA not associated with a living fish would likely have degraded). Thus we have especially high confidence that at least one live bighead carp was present just south of the O'Brien Lock and Dam (Figure 2). Furthermore, detections in the Calumet Sag Channel of bighead and silver carp eDNA at multiple locations on at least one sample date suggest strongly that multiple individual fish of both species were present in the Calumet Sag Channel. Although bighead carp (and probably silver carp) can move long distances quickly (e.g., one individual moved about 9 miles in a day), more typical movement is about 1 mile per day (Kolar et al. 2007). Thus movement of a single fish is an unlikely explanation for positive results on one day at two locations separated by about 10 miles in the Calumet Sag Channel (Figure 2). Overall our results indicate with very high confidence that at least a few live bighead and silver carp inhabited the Calumet Sag Channel and that multiple silver carp were present in the Calumet River.

We draw inferences from negative results (i.e., failure to detect eDNA) with considerably less confidence than from positive results because we know false negatives become more and more likely the lower the concentration of eDNA in the water. From sampling in the southerly pools, we know that even where target species are known to be present from traditional tools, we nevertheless did not detect eDNA in some samples. Thus, overall, negative results must be interpreted with great caution no matter. A negative result does not necessarily imply that no silver carp or bighead carp are present. It means only that the concentration of eDNA was lower than the detection limits of our current eDNA protocols.

Some have suggested that the following might provide alternative explanations for the presence of eDNA (alternative, that is, to the presence of at least one live fish): i) sewage treatment effluent from humans that had consumed silver or bighead carp or discarded fish waste, ii) deposition of excrement by seagulls or other birds that may have consumed silver or bighead carp tissue at other locations, iii) humans discarding one or more carcasses of silver or bighead carp directly into the canal, and iv) transport and release by barges of water containing eDNA. We consider each of these alternative explanations more fully below, and conclude that none of them is consistent with the overall spatial pattern of our positive results, especially those that have been shown up in repeated sampling trips (Figures 2).

Sewage treatment effluent is unlikely to contain eDNA from fish that were consumed or even from fish waste that may have been put down a kitchen sink because the DNA would degrade during passage through a sewage treatment plant even more quickly than it would degrade in the environment. In addition, the overall spatial pattern of positive results (Figure 2) is not consistent with sewage treatment outfall(s) as a source(s).

Excrement from birds (or humans) is unlikely to contain detectable quantities of DNA because the DNA would degrade substantially during passage through the digestive tract.

It is possible that humans may occasionally discard the waste from a cleaned silver or bighead carp caught or purchased elsewhere, and that DNA could thus be detected in the water in the immediate vicinity of a carcass. Live, fresh, and frozen bighead carp are commonly available for sale in Asian food markets in Chicago (and other major cities in the Great Lakes region). However, the geographic distribution of positive results and positive results in the same location on multiple dates suggest that discarded carcasses are an extremely unlikely general explanation for silver and bighead eDNA detections.

Transport of water used for ballast in barges could contain eDNA for silver and bighead carp if a barge took on water in carp-infested waters and discharged it as it traveled northward. However there is no barge traffic near the Wilmette Pumping Station, and U.S. Coast Guard guidelines to which the navigation industry agreed in September were supposed to have precluded ballast as an explanation in the Calumet Sag Channel, at least for our last sampling date (Figure 2).

Based on our understanding of the waterway and other potential pathways, we believe that no explanation other than the presence of multiple living silver and bighead carps can plausibly explain the entire repeated spatial pattern of positive results for silver and bighead carp eDNA in the waterway.

The repeated spatial pattern of our results indicate the both silver carp and bighead carp (Figure 2) have moved upstream and accumulated below barriers and other structures, both because passage is slowed or prevented, but also perhaps because those are favored habitats for these fishes, as indicated by some evidence from the Missouri River (Kolar et al. 2007).

How many carp will it take to launch an invasion in Lake Michigan?

Although our eDNA evidence suggests that at least one silver carp has swum into Lake Michigan, an "invasion" is not necessarily underway. In the most important sense of the word, "invasion" refers to an established and spreading population in a new location—one that is not only self-sustaining, but one that is growing. For that to happen, not only do at least one male and female fish have to survive to breeding season, they have to find each other at the right time and place to reproduce, and at least two of their young have to survive the gauntlet of predators and other hazards to themselves achieve adulthood and reproduce successfully.

And because each of those things happening is unlikely in any given circumstance, most incipient invasions fail (Lockwood 2005, Drake & Lodge 2006). Until the whole sequence of survival and reproduction happens many times, an incipient invading population remains highly vulnerable to extinction, even without any management efforts to eradicate them. In fact, even when fisheries managers, wildlife managers, biocontrol experts, or citizens have tried hard to start populations of (desirable) species, they usually fail. Most populations become established only after repeated introductions of large numbers of individuals. And many of those populations that initially failed multiple times eventually grew and spread rapidly. In summary, then, there is no evidence to suggest that a population of silver carp has established in Lake Michigan, and there is no evidence at all that bighead carp have entered Lake Michigan.

Given the vagaries of the survival and growth of a potentially invading population, no one knows how many silver carp or how many bighead carp it would take to launch a self-sustaining population of either species in Lake Michigan. However scientific studies do provide some useful guidance for management.

It is a numbers game. The more individual fish that swim into the lake, and the more times that happens, the more likely an invasion is to occur. Given a management goal to decrease the probability of an invasion of Lake Michigan, our scientific understanding—as well as common sense--gives us great confidence that the best way to accomplish the management goal is to reduce the probability of additional fish swimming into Lake Michigan.

Considering that invasion is *not* inevitable for either species, there is a high probability that management actions now that prevent additional silver and bighead carp from entering Lake Michigan could prevent population establishment in Lake Michigan. This is true even if some individuals of one or both species have already entered Lake Michigan. However the fact that one or more individual silver carp have apparently already entered Lake Michigan means that management efforts are more urgent than before we had that knowledge. Each time more fish enter the lake, we roll the dice to determine whether an invasion will result.

Compared to other pathways to Lake Michigan, how important is the Chicago canal?

If the management goal is to reduce the future number of individual silver and bighead carps gaining access to Lake Michigan, then it begs some questions: how did the fish get above the barrier? how might fish continue to gain access to the canal above the barrier? and which route is likely to give access to the greatest number of individual fish (and therefore should be the highest priority for management)? I tackle each of these questions in turn.

At least four possible explanations exist for how fish may have passed through or circumvented the electric barrier. First, the effectiveness of the electric barriers are unlikely to have been continuously 100% effective in the past and they are unlikely to be 100% effective in the future. Some silver and bighead carp may therefore have gone through the original barrier (or even both barriers in recent years), even while they were operational. Second, the barriers have failed occasionally. Third, in 2009, after barrier IIA was made operational, barrier I was temporarily shut down for maintenance. Although the Army Corps of Engineers did not believe that Asian carps were near the barriers at that time, it now seems more likely (given our eDNA results) that they were. Any Asian carp that had accumulated below barrier I (but above barrier IIA) would then have been free to swim upstream undetected and undeterred. Frequent observations of common carp behavior just below the barrier, and our eDNA results, suggest that such "stacking up" of fish just below the electric barriers and other upstream barriers is common.

Fourth, to keep the maps as clear as possible, we have not indicated sampling effort (Figure 1) or sampling results (Figure 2) in the Des Plaines River north of its confluence with the Chicago Sanitary and Ship Canal or in the I&M Canal. However, we have detected silver carp eDNA at a single location in the Des Plaines River about four miles north of the electric barriers (which are in the Chicago Sanitary and Ship Canal). In this area, the Des Plaines River runs parallel to the Chicago Sanitary and Ship Canal on the Canal's west side <100 meters from the Canal; the two are separated by a small elevation which is overtopped during high floods. Likewise, in the I&M Canal, which in the same region parallels the Chicago Sanitary and Ship Canal on the east side of the Canal, we detected eDNA for silver and bighead carp at multiple locations. The I&M canal connects with the CSSC above and below the electric barrier, and could possible provide a pathway to bypass the barrier. Thus, at times of high water, as in fall 2008, any Asian carps in either the Des Plaines River or the I&M Canal could have easily swum into the main canal above the electric barriers.

Given the four possibilities listed above, the presence of living silver and bighead carps in the canal north of the electric barriers is most plausibly explained by failures of the electric barrier to completely restrict the northward movement of silver and bighead carps.

However, it is also plausible that humans have released living silver or bighead carp individuals into parts of the waterway north of the electric barriers (Kolar et al. 2007). It is possible that juvenile silver and/or bighead carps have been unintentionally sold as live fish bait, most likely mixed with native fish species, some of which they resemble as juveniles. If so, some could have been released or escaped from anglers. However, because neither species of carp survives well under confinement like that in the bait industry, we do not see this vehicle as very likely.

Far more likely is the intentional release, at least in the past, of larger individuals, especially of bighead carp, purchased from Asian food markets (Shiu & Stokes 2008). The confirmed presence of silver and/or bighead carps in multiple Chicago area park ponds is strong evidence that intentional human release of these carps is sufficiently common for multiple individual carp to survive in multiple ponds and lakes. However, since 2003, it has been illegal in Chicago for silver and bighead carp to leave a food market alive. Because individual carp can live for many years, some individuals in the canal could have been released many years ago.

These considerations suggest at least three focuses of management action are appropriate. First, efforts to further minimize all the mechanisms by which additional fish might otherwise gain future access to the canal north of the electric barriers are essential. Successful management, and ideally eradication, of Asian carp above the barrier relies upon preventing further introductions. Although escape from or release by humans may have contributed to the populations of silver and bighead carps whose eDNA we have detected above the barrier, dispersal of Asian carps through or around the electric barriers is clearly the largest single potential source of future invaders.

Second, it is well established from traditional surveillance methods that the populations of both species are very large south of the barriers, and we know from a 30 year history in the Mississippi River and tributaries that both species disperse upstream whenever the opportunity exists. Once the populations of either or both species build up to very high numbers immediately south of the barriers, even a low failure rate of the barriers could result in a sufficient number of escapes north to cause an invasion in Lake Michigan unless additional management steps are taken. Suppression or control of the Asian carp population in the pools immediately below the barrier would help reduce the impact of catastrophic failure.

Third, efforts to decrease the survival and/or prevent the egress into Lake Michigan by those fish already in the canal north of the barrier is urgent. Toward that end, it makes no difference how the fish got there.

Overall, considering the most plausible pathways, dispersal northward of Asian carps that are currently south of the electric barriers poses the numerically greatest risk to the Great Lakes. It therefore merits the greatest management attention. I offer additional considerations for management in the last section of this testimony.

If one or both species of carps invade the Great Lakes, how likely are substantial damages?

A review of the potential damages to the Great Lakes environment and the regional economy is beyond the scope of my testimony. Three things however are clear. First, in many places with environments that are similar to the Great Lakes, silver carp and/or bighead carp have achieved spectacularly large abundances and caused declines of native fishes including close European relatives of our highly prized walleye (Kolar et al. 2007).

Second, it is therefore easy to imagine catastrophic results of an invasion by either silver carp or bighead carp (Newcomb 2010). Conversely, given the wealth of experience with both these species in many parts of the world, including temperate zone lakes and reservoirs, and including our own Mississippi and Illinois rivers, it is difficult to imagine either species having only a small impact. It is virtually impossible to imagine impacts that more than a very small number of people would consider to be positive.

Third, the experience elsewhere with these specieis suggests that impacts would be spread unevenly among the five Great Lakes. Lake Superior would probably be the least hospitable for both species, while both silver and bighead carps would probably thrive most in Lake Erie. One of the most valuable commercial and recreational freshwater fisheries in the world is at high risk (Rothlisberger et al. 2009).

Thus, while no one can predict the outcome with certainty if either silver carp or bighead carp establish in the Great Lakes, I believe the outcome is likely to be high environmental and economic damage if either invades, and worse if both invade. Whether the damage would be small or large, it would be irreversible for the foreseeable future because no acceptable technology currently exists to control their abundance in the Great Lakes.

In the context of endangered species, you've probably heard it said that "extinction is forever." Unfortunately, it is also usually true that invasion is forever. Biological invasions are the least reversible form of pollution. In contrast, most other forms of pollution--like the phosphorus pollution that we've been so successful at cleaning up in the Great Lake, the nitrogen and sulfur compounds of air pollution, heavy metals, and PCBs—degrade or get buried, and the problems they cause decline eventually, if only we stop adding molecules of them to the environment. Chemical pollutants, in other words, do not reproduce; species do. Once a population is established, even if we stop adding individuals to Lake Michigan, their populations will continue to grow, they will continue to spread throughout the Great Lakes, and their damage to fisheries, recreation, and human health will grow exponentially over time.

If no additional management and policy actions are taken, we would be playing Russian roulette with the Great Lakes environment and regional economy. And the gun with which we are playing has not just one but two chambers loaded with bullets. Even if silver carp has little impact, bighead carp might, or *vice-versa*. In fact, as I'll argue below, the Chicago canal has loaded many more chambers of our invasive species gun.

Does the Chicago canal pose a threat only from Asian carps? Only to the Great Lakes?

The issues before this committee should not be seen as limited to silver and bighead carp. They should also not be seen as limited to the Great Lakes and surrounding states.

The electric barriers that are now often referred to as the Asian carp barriers were originally motivated by an invasive fish going in the other direction—the round goby going from the Great Lakes (where it was introduced in ships' ballast) into the Mississippi River basin. By the time barrier I was operational, the round goby was several miles south of the barrier site. From the beginning, however, the barrier was seen even by its proponents as a stop-gap and partial solution to the threat for invasive species. There are many potentially harmful species that have invaded one but not yet both the Mississippi River basin and the Great Lakes basin. The most recent counts show about 186 nonindigenous species in the Great Lakes, most of which have not yet colonized the Mississippi River. These include the fish Eurasian River Ruffe, a fish likely to be quite harmful to the biota and fisheries of the Mississippi River basin, and small crustaceans (e.g., bloody red shrimp) that will not be slowed much by electricity. Less study has occurred in the Mississippi River basin, but there are many more potentially harmful species following the Asian carps northward, including another Asian carp (black carp), snakehead, and some highly invasive aquatic weeds that will not be slowed by electricity.

Thus the canal is a two-way highway to environmental and economic havoc, and benefits of any management steps that reduce the species traffic will extend far beyond silver and bighead carp. They will also bring benefits to the states in the Mississippi River basin as well as to the states bordering the Great Lakes. The invasive species gun with which we are playing Russian roulette has all its barrels loaded and in fact is primed to explode in both directions. Any management actions that make the passage of fishes through the canal harder, and any improvements that also reduce the potential for transit of additional organisms—those not affected by electricity--will bring long term additional benefits to North America, on top of the benefits derived from preventing invasion of the Great Lakes by one or more Asian carp species.

Science-based management actions for the Chicago canal

Given the goal shared by all federal agencies to prevent an invasion of the Great Lakes by either silver carp or bighead carp, any management action that reduces the likelihood of individuals of either species entering the lake should be seriously considered. First, while other options are considered, I recommend that urgent attention be given to any management action that will prevent the silver and bighead carp that are currently above the barrier from entering Lake Michigan. Second, options for eradicating or at least dramatically reducing the numbers of the individuals above the barrier should also be considered. Third, the operation and maintenance of the two existing barriers, and the plans for the third barrier, should be fine-tuned as much as possible to maximize effectiveness against fishes moving in either direction (barrier IIA was designed to be more effective against species moving northward). Fourth, a surveillance program needs to be established in the Great Lakes to locate and determine the extent of any Asian carp presence in the Great Lakes, targeted perhaps at the tributaries most likely to support spawning of the carps. This should be coupled with development of methods that would allow any fish detected to be contained, and eradicated. Fifth, other deterrents to fish movement should be considered to augment the barriers. Sixth, the Mississippi River basin and the Great Lakes basin should be permanently separated ecologically, as agreed among many agencies, stakeholders and experts at the 2003 canal summit in Chicago (Brammier et al. 2008). It is not only Asian carp we should be thinking about, but the hundreds of potentially harmful species (many of them completely unaffected by electrical current) in both basins, the damages from which would be suffered by us and our children in perpetuity.

Finally, it is important to keep in mind that while we talk, the Asian carps swim. They do not respect our timetable, our budgets, or our political boundaries. And there are many more species on each side of the canal that also will not wait for our management deliberations before they too traverse the canal.

REFERENCES CITED

[1] Brammieier, J., Polls, I. & Mackey, S. (2008). *Preliminary feasibility of ecological separation of the Mississippi River and the Great Lakes to prevent the transfer of aquatic invasive species*. Great Lakes Fishery Commission project completion report.

[2] Drake, J. M. & Lodge, D. M. (2006). Allee effects, propagule pressure and the probability of establishment: risk analysis for biological invasions. *Biological Invasions, 8*, 365-375.

[3] Ficetola, G. F., Miaud, C., Pompanon, F. & Taberlet, P. (2008). Species detection using environmental DNA from water samples. *Biology Letters, 4*, 423-425.

[4] Fowler, A. J., Lodge, D. M. & Hsia, J. F. (2007). Failure of the Lacey Act to protect U.S. ecosystems against animal invasions. *Frontiers of Ecology and Environment, 5*, 353-359.

[5] Fischer, J. R. & Paukert, C. P. (2009). Effects of sampling effort,

assemblage similarity, and habitat heterogeneity on estimates of species richness and relative abundance of stream fishes. *Canadian Journal of Fisheries and Aquatic Sciences, 66*, 277-290.

[6] Keller, R. P., Lodge, D. M., Lewis, M. & Shogren, J. F. (eds). (2009). *Bioeconomics of invasive species: integrating ecology, economics, policy and management*. Oxford University Press.

[7] Kerr, S. J., Brousseau, C. S. & Muschett, M. (2005). Invasive aquatic species in Ontario: a review and analysis of potential pathways for introduction. *Fisheries, 30*, 21-30.

[8] Kolar, C. S., Chapman, D. C., Courtenay, Jr., W. R., Housel, C. M., Williams, J. D. & Jennings, D. P. (2007). Bigheaded carps: a biological synopsis and environmental risk assessment. American Fisheries Society, Special Publication 33, Bethesda, Maryland.

[9] Lockwood, J. L., Cassey, P. & Blackburn, T. (2005). The role of propagule pressure in explaining species invasions. *Trends in Ecology and Environment, 5*, 223-228.

[10] Lodge, D. M., Williams, S., MacIsaac, H., Hayes, K., Leung, B., Reichard, S., Mack, R. N., Moyle, P. B., Smith, M., Andow, D. A., Carlton, J. T. & McMichael, A. (2006). Biological invasions: recommendations for U.S. policy and management. *Ecological Applications, 16*, 2035-2054.

[11] Lodge, D. M. (2009). Declaration in Appendix to United States Memorandum in Opposition. http://www.supremecourtus.gov/ SpecMastRpt/ ApptoMemfortheUSinOpposition.pdf

[12] Newcomb, T. J. Affidavit in Appendix to Motion for Preliminary Injunction. http://www.supremecourtus.gov/SpecMastRpt/Orig%201,%202%20&% 203%20Appendi x.pdf

[13] Magnuson, J. J., Benson, B. J. & McLain, A. S. (1994). Insights on species richness and turnover from long-term ecological research: fishes in north temperate lakes. *American Zoologist, 34*, 437-451.

[14] Rothlisberger, J. D., Lodge, D. M., Cooke, R. M. & Finnoff, D. C. (2009). Future declines of the binational Laurentian Great Lakes fisheries: recognizing the importance of environmental and cultural change. Frontiers in Ecology and Environment (in press) doi: 10.1890/090002.

[15] Shiu, H. & Stokes, L. (2008). Buddhist animal release practices: historic, environmental, public health and economic concerns. *Contemporary Buddhism, 9*, 181-196.

In: Troubled Waters: Asian Carp and the… ISBN: 978-1-61761-766-9
Editors: Deborah M. Maeda

Chapter 9

TESTIMONY OF DR. MICHAEL J. HANSEN, CHAIR, GREAT LAKES FISHERY COMMISSION, BEFORE THE SUBCOMMITTEE ON WATER RESOURCES AND ENVIRONMENT, HEARING ON "ASIAN CARP AND THE GREAT LAKES"

INVASIVE SPECIES AND THE DESTRUCTION THEY BRING

Madam Chair, thank you for inviting me to appear before this subcommittee to discuss the threat of the Asian carp invasion into the Great Lakes. My name is Michael Hansen. I am the chair of the Great Lakes Fishery Commission. I am also a professor of fisheries at the University of Wisconsin at Stevens Point.

The Great Lakes are an extremely valuable resource for both the United States and Canada. The Great Lakes' commercial, recreational, and tribal fisheries are valued at more than $7 billion annually (ASA 2008). The lakes provide drinking water for 40 million people and are a rich tourist draw. They are a way of life for the people of the region and a healthy, vibrant Great Lakes ecosystem is immeasurable in economic terms alone.

The Great Lakes—and the way of life they support—are under assault from invasive species. Invasive species are defined as non-native animals and plants, both aquatic and terrestrial, that enter new environments, become established, and spread. The Great Lakes are "ground zero" for aquatic invasions. Today, the lakes harbor more than 185 non-native species (Lodge 2007; Mills et al. 1993; Ricciardi 2001; Sturtevant et al. 2010), many of which entered the lakes accidentally. The rate of introduction into the Great Lakes is not slowing, even with the welcomed institution of some invasive species control measures (e.g., ballast water exchange requirements starting as early as 1989). Some estimate that a new invader enters the system every 9-12 months. Many in the scientific community, however, believe that the Great Lakes contain many more invasive species than have been discovered, because a coordinated, basinwide program to monitor new nonindigenous species does not exist (IAGLR 2008; Sturtevant et al. 2010).

Invasive species have many pathways into new ecosystems. Ballast water is a major vector, as are canals and waterways, the trade of live organisms, recreational activities, and aquaculture. While much of the focus has been on large or prominent organisms, microorganisms and pathogens are also an increasing concern (particularly with the emergence of the VHS virus, an exotic fish disease linked to fish kills in several Great Lakes and just recently detected in Lake Superior). The Great Lakes, essentially, are a welcoming, open door for invaders.

The world and North America are becoming more globalized. With enhanced trade and the movement of goods comes the reality that more species have more pathways than ever to invade the Great Lakes. The Saint Lawrence Seaway, for instance, is a direct pathway for foreign ships into the U.S. heartland. Those ocean going ships have been responsible for more than 1/3 of all Great Lakes invaders (Mills et al. 1993; Sturtevant et al. 2010). Also, the U.S. Fish and Wildlife Service reports that an average of more than 200 million fish, and tens of millions of reptiles and amphibians, birds, and mammals are imported into the United States annually for food and for the pet trade industry. Fish for pet trade, for example, are collected in exotic locations throughout the world or reared in aquaculture facilities (Livengood and Chapman 2007), which are prone to flooding, thereby enabling escapement.

Invasive species are not a local or even a regional problem—they are a national and a global problem. Invasive species spread readily from region to region, so species introduced into one part of the country will, in all likelihood, eventually make it to other parts of the country. Eurasian *Dreissenid* mussels (a.k.a., the "zebra mussel"), for instance, entered the Great Lakes through ballast water from oceanic ships in the mid-1980s and have now spread throughout much of the United States. Asian carp, which are discussed below, escaped from aquaculture in the Deep South and, as they made their way northward through the Mississippi and Illinois Rivers, have become a major economic and ecological nuisance. These carp are now found in Texas, the Ohio River Basin, and are threatening the Great Lakes and even the Columbia River Basin in the Pacific Northwest. Snakeheads were imported for the aquarium trade and for food and are now present in the Northeast, the East, and the Mississippi River system. Specimens have also been found in Alabama, California, Florida, Kentucky, Texas, Washington, and Lake Michigan. The point is, exotic introductions into United States' waters anywhere raise the possibility of spread to other ecosystems. Solutions must be large in scope and based on the assumption that species will multiply and extend their range.

The focus of this hearing is on Asian carp, and the primary pathway for Asian carp to enter the Great Lakes from the Mississippi River basin is through two canals in the Chicago area: the Chicago Sanitary and Ship Canal and the Cal-Sag Canal. Other witnesses during this hearing will address the policies that are being undertaken to try to stop the spread of Asian carp through those canals. I will reflect on these policies toward the end of this statement. Let me first provide a summary of the threat Asian carp pose to the Great Lakes.

THE HAVOC OF ASIAN CARP: A HARBINGER OF WHAT'S TO COME FOR THE GREAT LAKES?

Asian carp have the ability to spread rapidly, reproduce in large numbers, and become the predominant species in an ecosystem. Once established, fishery managers have little chance to control the fish. Like the sea lamprey, they could become a permanent element of the Great Lakes if they enter the system.

The term "Asian carp" is a generic term to describe several species of related fish originating from Asia. Two species of Asian carp primarily comprise the current invasionvia the Illinois Waterway System —the "bighead" and "silver" carps. These species were imported into the southern United States to keep aquaculture facilities clean and to serve the food fish industry. Bighead carp were imported from China in 1972. A year later, in 1973, silver carp were brought into the United States from China and eastern Siberia(Chick and Pegg 2001; Hoff 2008; Schrank and Guy 2002; Tucker et al. 1996) . By 1980, bighead and silver carps, which had escaped from aquaculture facilities, had been captured in the wild by fishers in Arkansas,

Louisiana, and Kentucky (Williamson and Garvey 2005). Flooding events in the 1980s and 1990s allowed the bighead and silver carps to greatly expand their range. The floods provided extensive spawning and rearing habitat that facilitated high survival rates for offspring.

Since their escape nearly two decades ago, bighead and silver carps have overwhelmed the Mississippi and Illinois River systems. Bighead and silver carp are filter feeders. They eat plankton (e.g., algae and microscopic animals), the very foundation of the food web. Their feeding habits were the reason they were imported into the United States by the aquaculture industry: by feasting on plankton, they kept aquaculture facilities clean. Nevertheless, when loose in the wild, where plankton were abundant and predators were few, the Asian carp had a field day. Between 1991 and 2000, as the invasion was unfolding, biologists observed an exponential increase in bighead carp numbers in the Illinois River, near St. Louis (Chick and Pegg 2001) . Such increases played out time and again as the carp expanded their range northward. Commercial harvest of bighead carp in the Mississippi River Basin, for instance, increased from 5.5 tons to 55 tons between 1994 and 1997 (Chick and Pegg 2001). Biologists reported dietary overlap among Asian carp and native fishes in the Mississippi and Illinois Rivers, which suggests the Asian carp would likely outcompete native fish for food. In fall 1999, an investigation of a fish kill in off-channel waters of a National Wildlife Refuge near St. Louis documented that Asian carp made up 97% of the biomass (MICRA 2002), which indicates that, at least in that area, the fish community consisted of almost nothing but Asian carp. During this period, commercial fisherman began reporting that they were abandoning their traditional fishing sites because they were unable to lift nets that were "loaded" with Asian carp. Today, commercial fishers in the Illinois River regularly catch upwards of 25,000 pounds (11,000 kg) of bighead and silver carp *per day* (Irons et al. 2007). A half of an acre can often yield thousands of pounds of Asian carp (Chapman 2003), an astonishing amount of fish and an indicator of just how much of total fish biomass Asian carp can represent. The commercial value of Asian carp is extremely lowand much less valuable than the native fish they replaced .

Biologists and policy makers are particularly troubled by the fact that Asian cap can grow to extremely large size because an Asian carp is capable of eating 40% of its body weight each day (Hoff 2004). Bighead and silver carp voraciously consume plankton, stripping the food web of the key source of food for small and big fish.

The silver variety of the Asian carp has a unique characteristic that makes it particularly dangerous to humans: the sound of a boat motor startles the fish, causing it to leap as high as ten feet out of the water. These flying fish—some weighing more than twenty pounds—serve as a projectile, landing in boats, damaging property, and injuring people. Biologists on the Illinois River need to follow new safety protocols to avoid serious injuries from these fish. Waterskiing and other aquatic activities have grown extremely dangerous. The newspapers and YouTube are replete with accounts of people being injured by Asian carp, including a story about woman who nearly died in 2004 after being knocked unconscious from her Jet Ski near Peoria, Illinois (Meersman 2004). Said Duane Chapman of the U.S. Geological Survey, a biologists in the thick of these fish, "You may imagine it would be quite novel for a 20-pound fish to jump into your boat, but being hit by a large Asian carp would be similar to being hit by a bowling ball. Even if the fish don't hit you, they can break fishing rods, windshields, electronics or anything else in your boat. As if adding insult, the carp will leave slime, blood and excrement on everything it touches" (Chapman 2010). The public's safety and property are clearly at risk.

The trail of destruction—to the ecosystem, economy, property, and boaters—that these Asian carp have left in their wake has been cause for tremendous concern to the people of the Great Lakes basin. Would Asian carp have a similar impact on the Great Lakes basin as theydid in the Mississippi and Illinois River systems? We will have littlechance of managing these newfish es if they become establishedin the Great Lakes.

Risk assessments carried out by officials from the U.S. Department of Interior (Kolar et al. 2005) and the Department of Fisheries and Oceans Canada (Mandrak and Cudmore 2004), and overall experience with biological invasions, give little reason to be optimistic. For starters, these assessments indicate that the carp are certain to tolerate the Great Lakes basin's climate, because the basin is well within the fishes' native climate range. Mean annual air temperatures range between -2°C and 22°C for bighead carp and -6°C and 24°C for silver carp, a temperature span that would support Asian carp populations in much of the United States and Canada, including the Great Lakes.

Risk assessments also indicate that the carp would likely find the Great Lakes to contain an abundant and diverse supply of food. In the Great Lakes, the bighead carp would consume zooplankton and silver carp would prey heavily on phytoplankton, thereby competing with the young of many native species and all life stages of native planktivorous fish species. To make matters worse, Asian carp do not appear to be too finicky about what they eat. For instance, bighead carp diet in the Mississippi River is more varied than in theirnative range, because they feed on algae, detritus, and zooplankton. This means that the carp appear to be able to feed opportunistically. Also, by feeding on plankton, the Asian carp feed on the "low end" of the food web. That is, they will compete for food with the young of many native fish species and with all life stages of planktivorous native fish. Little doubt exists that bighead and silver carp would have significant negative impacts on the food web by causing large-scale changes at the low end of the structure.

The Asian carp need certain types of habitat to feed and spawn successfully, including tributaries greater than 30 miles (50 km) of unimpeded length. The carp would also thrive in areas with vegetated shorelines that afford them suitable habitat for feeding. The Great Lakes basin contains numerous streams with suitable spawning habitat and large areas of vegetated shorelines, particularly large bays, wide river mouths, connecting channels (e.g., the Saint Marys River), wetlands, and lentic areas (areas of still waters). While the carp may not thrive in large portions of the basin—for example, in the deep, cold, open waters of the lakes—all lakes, including Lake Superior, contain ample habitat for spawning and feeding.

Should the silver carp become established in the Great Lakes basin, they will likely inflict harm directly on people. The Great Lakes Commission estimates that nearly 1 million boats and personal watercraft operate on the lakes (GLC 2003), which thereby places millions of people in potential contact with the silver carp. Knowing the hazards of boating, Jet-skiing, and waterskiing on the Illinois River system, the problem of projectile fish would be compounded on the Great Lakes by a significantly larger boating population in the region.

Overall, citizens of the Great Lakes region should be deeply concerned about the prospects of Asian carp. Mandrak and Cudmore (2004) concluded that the probability of bighead and silver carps surviving and reproducing in the Great Lakes is high. If bighead and silver carp colonize the Great Lakes, they will likely spread throughout the basin due to the natural and man-made connections and the widespread distribution of suitable habitat.

Policy Issues

The history of aquatic invasions has shown that people are left with few options to control a species once the species enters an ecosystem and spreads. With sea lampreys, the region has been relatively fortunate in that the species concentrates in streams and is vulnerable to control during several portions of its life cycle. Also, the alewife, while a nuisance, serves as a food fish for predators like trout and salmon, thereby making that species controllable through stocking and rehabilitation programs. Other than those two species, meaningful control mechanisms do not exist in the Great Lakes basin for other invaders.

The short answer to the question “What can be done if Asian carp enter the Great Lakes?” is “Not much.” At least, not much at the moment. Currently, control mechanisms do not exist for Asian carp, if they become established. Scientists do not know of a pesticide that would target the carp, nor weaknesses in their spawning behavior that could be exploited, nor predatory pressures that would help reduce populations. That said, the effort to find solutions has not been robust. The sea lamprey control program has been a success because of a concerted effort to apply science to discover control techniques. Sea lamprey control has worked because lines of accountability are clear—the Great Lakes Fishery Commission is responsible. Sea lamprey control has worked because the governments of Canada and the United States have committed resources to do the job. Currently, no such effort exists for other invasive species, including Asian carp. Granted, universities and government agencies are conducting solid, promising research on invasive species, but until governments redouble their efforts—both in terms of resources and in terms of vision—viable solutions for any invasive species are probably decades away. The Great Lakes Regional Collaboration’s Aquatic Invasive Species Team noted as much and recommended the establishment of an “Integrated Pest Management Program” to focus attention of government. The commission strongly agrees and recommends a concerted effort to find solutions to some of the most pressing invasive species problems. Such solutions include both the development of control techniques and the establishment of accountability so that an agency remains motivated toward progress.

This paucity of control options has been a strong force motivating prevention. The Great Lakes Fishery Commission has been a partner with other primary agencies to seek preventative measures for Asian carp for more than a decade. These measures were discussed in greater detail by other panelists during today’s hearing. The commission has joined its partners over the years in pressing for construction of an electrical dispersal barrier, stopping trade of live Asian carp, and supporting other steps taken by management agencies in the Chicago region. The commission strongly supports current efforts to complete the electrical barrier, to build a structure of some kind to prevent species transfer between rivers that parallel the Chicago Sanitary and

Ship Canal, and to plug other holes (such as culverts and pipes) that might allow species migration. The commission is heartened by the strong interest that Cameron Davis, the Senior Advisor to the EPA Administrator, has taken in this issue, because the administration's interest in coordinating a multi-agency response isbad ly needed.

While current work to prevent Asian carp migration are certainly appropriate, the only solution to this problem is to achieve what is called "ecological separation," that is, altering the canal system in a way where it is *impossible* for species of any kind to move from the Mississippi basin to the Great Lakes or vice versa. This separation was included as a recommendation of the Aquatic Invasive Species Summit convened by Chicago Mayor Richard M. Daley in 2003 (Anonymous 2003).

The recommendation from 2003 was to achieve that separation "within 10 years," so much needs to be done in a short amount of time. In fact, the Great Lakes do not have any time to lose. Ecological separation must occur immediately. To kick-start the investigation into the feasibility of ecological separation, the Great Lakes Fishery Commission and the Great Lakes Fishery Trust co-commissioned a study to examine transportation patterns on the waterways, the hydrology, and options for achieving separation. That report (Brammeier et al. 2008) was completed about a year ago and its conclusions have never been more relevant. The commission appreciates Mr. Brammeier and his co-author's work on this issue and thanks the chair for including him at this hearing, because his insights are critical to understanding ultimate policy solutions.

Finally, the commission recognizes that the Brammeier project is really the start to a serious look at achieving ecological separation. The *Water Resources Development Act of 2007* authorized the U.S. Army Corps of Engineers to conduct a full-scale engineering analysis to identify and propose ways to achieve this essential separation. The commission was gratified to see the corps begin this study during the first fiscal year after it was authorized. The corps will continue with the study in 2010. The commission urges the corps to complete this study with all haste. The Great Lakes Fishery Commission urges Congress to clearly express that the end objective is ecological separation—not "reduce the risk" or "try to achieve separation while maintaining the status quo," the goal must be "ecological separation." Further, the commission urges Congress to provide the corps with resources to accelerate development of solutions that will achieve ecological separation and for Congress, at this time, to provide the corps with the authority it needs to implement any solution it proposes, so long as the solution fully achieves the separation goal. The Great Lakes Fishery Commission is concerned that the corps' study will be protracted and that separation will be delayed as authorizations and appropriations for a recommended project wind their way through the legislative process. The Great Lakes cannot wait.

LESSONS FROM THE SEA LAMPREY

Before I conclude, I would like to emphasize why prevention is paramount and why all efforts to address Asian carp have been essential. The Great Lakes Fishery Commission, the organization I chair, knows a great deal about invasive species. The commission was established in 1955 by the Canadian and U.S. *Convention on Great Lakes Fisheries* (U.S. Department of State 1956), primarily as a response to one of the most injurious invaders to ever enter the Great Lakes system: the sea lamprey.

Sea lampreys are primitive eel-like fishes native to the Atlantic Ocean. Shipping canals were the primary vector for sea lampreys to invade the upper Great Lakes in the early 1900s through improvements to the Welland Canal, which was built to bypass Niagara Falls. Sea lampreys are parasites in their native environment, but were able to wreak staggering damage on the Great Lakes ecosystem. By the late 1940s, harvest of lake trout, a keystone species, had fallen by 99% from the average catch of the 1930s (Fetterolf and Krueger 1990). The fishery that once sustained native fishers, fueled lucrative commercial operations, and attracted millions of anglers who simply enjoyed the outdoors was devastated. In short, sea lampreys changed the human way of life in the Great Lakes basin. The problem was so great that the governments of Canada and the United States were largely motivated by the sea lamprey's devastation when they agreed to the *Convention on Great Lakes Fisheries*, and included sea lamprey control commitments in the treaty.

Since 1955 when the commission was formed, the commission has delivered a sea lamprey control program, in cooperation with the Department of Fisheries and Oceans Canada, the U.S. Fish and Wildlife Service, the U.S. Geological Survey, and the U.S. Army Corps of Engineers. The commission's control program successfully reduced sea lamprey populations by 90% in most areas of the Great Lakes. Nevertheless, eradication isimpossible and the ongoing program is expensive.

The sea lamprey has taught some tough lessons, which we would be well-served to heed as we consider the Asian carp threat:

- A single invasive species can cause significant, permanent damage to the economic and ecological health of a region. We are fortunate that sea lampreys can be controlled, but sea lampreys remain a permanent, destructive element in the Great Lakes basin. Most—if not all—fishery management decisions made by federal, state, tribal, and provincial agencies must forever account for sea lampreys.
- Control of invasive species, if possible, is expensive and ongoing. The commission has spent more than $300 million since 1956 controlling sea lampreys. This amount, while large, does not account for the billions of dollars of revenue lost to commercial, tribal, and recreational fishers of the Great Lakes basin, nor does it account for billions of dollars spent by state and federal governments over decades to rehabilitate and propagate the fishery after sea lamprey invasion. Moreover, this figure does not include the immeasurable damage to the ecology of the Great Lakes basin.
- **Prevention is key; eradication is not possible.** The Great Lakes fishery will forever contend with sea lampreys.
- Citizens shoulder the costs and consequences of invasive species, not the beneficiaries of open waterways for shipping, fish ponds for aquaculture, or the free trade of live organisms.
- Programs to manage invasive species are costly and borne by taxpayers.

Sea lampreys have taught us that prevention of new invaders is critical. Once a species enters an ecosystem and becomes established, few tools, if any, exist to manage, let alone eradicate, invasive species. In fact, of the more than 180 non-native species in the Great Lakes, sea lampreys and alewives are the only aquatic invasive species that are being managed.

What remains unclear is whether policy makers truly understand the sea lamprey's lesson. Even with all that is known about the damage of invasive species, and even though pathways are generally identified, precious little has been done to prevent new introductions. A meaningful process does not exist to assess the risk of proposed importations of live organisms or to discover ways to manage the harmful species that have become established. Myriad canals and artificial connections exist between naturally distinct watersheds, leaving the Great Lakes region vulnerable to invasions from other parts of the United States and, in turn, being a source of invaders. Ballast water regulations have been proposed but they are far from accepted or implemented.

CONCLUSION

Efforts to prevent the migration of Asian carp into the Great Lakes have been motivated by what has been observed in the Mississippi and Illinois River systems—large-scale ecosystem disruption, loss of once-viable commercial fisheries, and human harm. Risk assessments conclude that the Great Lakes would likely be suitable habitat for Asian carp. Because control techniques for Asian carp are non-existent, agencies have been working non-stop for years to create barriers on the Chicago Sanitary and Ship Canal, to stop the trade of live Asian carp, and to fill all known policy gaps. The job is far from complete. The only true solution is achieving ecological separation. With the administration's strong interest in coordinating the response, the Great Lakes Fishery Commission remains confident that such separation will occur as soon as possible. Madam Chair, thank you for holding this hearing and for any action the committee is willing to take to help us and the administration in its efforts.

REFERENCES

Anonymous. (2003). *Aquatic invasive species summit proceedings*. City of Chicago, Chicago. ASA. 2008. Today's angler: A statistical profile of anglers, their targeted species and expenditures. American Sportfishing Association, Alexandria, VA.

Brammeier, J., Polls, I. & Mackey, S. (2008). *Preliminary feasibility of ecological separation of the Mississippi River and the Great Lakes to prevent the transfer of aquatic invasive species*. Great Lakes Fishery Commission and Great Lakes Fishery Trust, Ann Arbor and Lansing.

Chapman, D. (2003). Bighead and silver carp in the Mississippi and Missouri Rivers, as viewed online on January 30, 2010 at infolink.cr.usgs.go v/Science/Documents/invasive_carp.pdf. United States Geological Survey, Columbia, MO.

Chapman, D. (2010). *Carp lemonade: Making the best out of some big-headed invaders*. Missouri Conservationist Online As accessed January 30, 2010 at mdc.mo.gov/conmag/2004/07/20.htm.

Chick, J. H., & Pegg, M. A. (2001). Invasive carp in the Mississippi River basin. *Science*, *292*(5525), 2250-2251.

Fetterolf, C. M. & Krueger, C. C. (1990). Statement on H.R. 4299, The Great

Lakes Fish and Wildlife Restoration Act of 1990. Great Lakes Fish and Wildlife Restoration Act of 1990: Hearing before the Subcommittee on Fisheries and Wildlife Conservation and the Environment, House Committee on Merchant Marine and Fisheries [Serial No. 101-111]. U.S. Government Printing Office, Washington.

GLC. (2003). Great Lakes recreational boating's economic punch: Pages *in* Great Lakes Commission, editor., Ann Arbor.

Hoff, M. (2004). Asian carp: Huge fish with huge impacts, as accessed online on January 30, 2010 at www.asiancarp.org/Documents/AsianCarp.pdf. United States Fish and Wildlife Service, Fort Snelling, MN.

Hoff, M. (2008). Personal communication, June, 2008.

IAGLR. (2008). Research and Management Priorities for Aquatic Invasive Species in the Great Lakes, at *http://iaglr.org/ scipolicy/ais/ backgrou*nd. php. International Association for Great Lakes Research. Accessed June 20, 2008.

Irons, K. S., Sass, G. G., McClelland, M. A. & Stafford, J. D. (2007). Reduced condition factor of two native fish species coincident with invasion of non-native Asian carps in the Illinois River, U.S.A. Is this evidence for competition and reduced fitness? *Journal of Fish Biology*, 71(Supplement D), 258-273.

Kolar, C. S. & coauthors. (2005). Asian carps of the genus *Hypophthalmichthys* (pisces, cyprinidae)--A biological synopsis and environmental risk assessment, as accessed on January 30, 2010 at www.fws.gov/contaminants/OtherDocuments/ACBSRAFinalReport2005. pdf. United States Fish and Wildlife Service, Washington, DC.

Livengood, E. J. & Chapman, F. A. (2007). The ornamental fish trade: *An introduction with perspectives for responsible aquarium fish ownership.* University of Florida IFAS Extension.

Lodge, D. M. (2007). Testimony by David M. Lodge, University of Notre Dame, September 27, 2007. Subcommittee on Fisheries, Wildlife, and Oceans, Committee on Natural Resources.

Mandrak, N. E. & Cudmore, B. (2004). Risk assessment for Asian carps in Canada, as accessed January 30, 2010 *www.dfo-mpo.gc.ca/csas/* Csas/DocREC/2004/RES2004_103_E.pdf. Fisheries and Oceans Canada, Great Lakes Laboratory for Fisheries and Aquatic Sciences, Burlington, ON.

Meersman, T. (2004). No one wins in a crash with a fish (June 15, 2004). Minneapolis-St. Paul Star Tribune.

MICRA. (2002). Asian carp threat to the Great Lakes. River Crossings: *The*

Newsletter of the Mississippi Interstate Cooperative Resource Association, 11(3), 1-2.

Mills, E. L., Leach, J. H., Carlton, J. & Secor, C. (1993). Exotic species in the Great Lakes: A history of biotic crises and anthropogenic introductions. *Journal of Great Lakes Research, 19(1),* 1-54.

Ricciardi, A. (2001). Facilitative interactions among aquatic invaders: Is an 'invasional meltdown' occurring in the Great Lakes? *Canadian Journal of Fisheries and Aquatic Science, 58*, 2513-2525.

Schrank, S. J. & Guy, C. S. (2002). Age, growth, and gonadal characteristics of adult bighead carp, *Hypophthalmichthys nobilis*, in the lower Missouri River. *Environmental Biology of Fishes, 64*, 443-450.

Sturtevant, R., Reid, D. F., Ricciardi, A., Kipp, R. & Fuller, P. (2010). Great Lakes aquatic nonindigenous species list, online at *http://www.glerl.n* oaa.gov/res/Programs/ncrais/great_lakes_list.html. National Center for Research on Aquatic Invasive Species, National Oceanic Atmospheric Administration, accessed June 23, 2008.

Tucker, J. K., Cronin, F. A., Hrabik, R. A., Petersen, M. D. & Herzog, D. P. (1996). The bighead carp (*Hypophthalmichthys nobilis*) in the Mississippi River. *Journal of Freshwater Ecology, 11*(2), 241-243.

U.S. Department of State. (1956). Convention on Great Lakes Fisheries between the United States of American and Canada, TIAS 3326. *Pages 2836-2842 in United States treaties and other international agreements,* Vol. *6*, part 3, volume 6, part 3. U.S. Government Printing Office, Washington.

Williamson, C. J. & Garvey, J. E. (2005). Growth, fecundity, and diets of newly established silver carp in the Middle Mississippi River. *Transactions of the American Fisheries Society, 134,* 1423-1430.

In: Troubled Waters: Asian Carp and the... ISBN: 978-1-61761-766-9
Editors: Deborah M. Maeda

Chapter 10

STATEMENT OF DEL WILKINS, VICE PRESIDENT OF TERMINAL OPERATIONS AND BUSINESS DEVELOPMENT, CANAL BARGE COMPANY, THE AMERICAN WATERWAYS OPERATORS, BEFORE THE SUBCOMMITTEE ON WATER RESOURCES AND ENVIRONMENT, HEARING ON "ASIAN CARP AND THE GREAT LAKES"

Good afternoon Madame Chairman. I am Del Wilkins, Vice President of Terminal Operations & Business Development for Canal Barge Company. Canal Barge Company is a family-owned business headquartered in New Orleans, Louisiana that has been in business for 76 years, and operates throughout the inland waterways system, from the Texas coast up to Pittsburgh and Chicago. Canal Barge owns Illinois Marine Towing, a Chicago-area towing and barge fleeting company, and Canal Terminal Company, a bulk liquid storage facility just south of Joliet, Illinois. Canal Barge and its affiliates are daily users of the Illinois waterways and bring both liquid and dry cargo in and out of the Chicago area from all over the country. I am pleased to testify today on behalf of the American Waterways Operators, the national trade association for the tugboat, towboat and barge industry. Thank you for the opportunity to be here and share our concerns about the manner in which government addresses the serious issue of the spread of Asian carp into the Great lakes.

As commercial users of the inland rivers, coastal waterways and Great Lakes, the tugboat, towboat and barge industry has a genuine commitment and obligation to stewardship of the environment in which we operate. Our industry is safe, environmentally responsible, cost-efficient, and uncommonly progressive in partnering with government to make our nation's waterways as safe and as clean as possible. Our full commitment to collaboration is best exemplified by the results of the more than 15 years of industry effort, in cooperation with Congress and the U.S. Coast Guard, to put into practice our commitment to safe operations and zero pollution. In 1994, AWO became the first transportation trade association to establish a code of safety and environmental stewardship for member companies. For a decade, the AWO Responsible Carrier Program has been a condition of membership in AWO, and members must pass an independent safety audit every three years or forfeit their membership in the association. Since 2004, AWO has cooperated with the Army Corps of Engineers, the Coast Guard, the Environmental Protection Agency and federal and state agencies concerning the operation of the electric fish barrier on the Chicago Sanitary and Ship Canal to protect the Great Lakes and the Western Rivers from invasive species, promoted the recovery of threatened and endangered species and established practices to reduce emissions from tank barges.

In the fight against the spread of Asian carp, and other invasive species, our fundamental message is this: the question of whether to protect the environment or ensure the continued flow of vital maritime commerce is an unnecessary choice, and one our nation cannot afford to make. We fully support robust protections for the Great Lakes from the Asian carp. We must also preserve human health and safety and maintain the free flow of waterborne commodities that are critical to the national and regional economy. We are confident that congressional and Administration leadership and stakeholder cooperation will lead us to a sustainable solution that protects the Great Lakes ecosystem without sacrificing critical jobs and the environmental and economic benefits of barge transportation.

Finding such a solution is critical because inland waterways navigation is essential to our economy and is the most environmentally friendly mode of freight transportation. A 2007 study conducted by the Texas Transportation Institute for the U.S. Department of Transportation found that inland waterways transportation is the safest, cleanest and most economical method of freight transportation. In a modal comparison, the study found that waterways transportation has the lowest hazardous material spill record and generates fewer emissions than rail or truck transportation. Moreover, a typical inland barge has a dry cargo capacity at least 16 times greater than a single rail car, and 70 times greater than a single semi-trailer truck. A single tank barge can carry as much liquid cargo as 144 trucks or 46 rail cars. Barge transportation plays an integral role in the multimodal freight transportation system of the Midwest-Great Lakes region by reducing truck and rail traffic and highway congestion.

Balanced solutions to the problem of invasive species in the Great Lakes are achievable. An integrated approach can arrest the advance of the Asian carp, protect the Great Lakes ecosystem, utilize science to develop more effective ways to combat the carp's spread, and maintain safe, efficient and reliable navigation on vital commercial waterways. We stand ready to act as a constructive partner with federal, state and other concerned stakeholders in that process. Below is a list of nine promising measures that warrant consideration and can be part an integrated strategy to prevent the spread of Asian carp.

- **First, expedite construction of the third barrier on the Chicago Sanitary and Ship Canal, known as Barrier IIB.** AWO continues to work with the Corps of Engineers and other stakeholders to ensure that electric fish barriers are both effective in preventing Asian carp movements while allowing for the safe passage of vessels and their crews. We support the immediate completion of Barrier IIB, coupled with cumulative safety tests of the voltage for all three barriers to ensure safe transit of commercial and recreational vessels.
- **Second, consider other types of fish barriers to prevent Asian carp from moving into the Lakes, such as bubble or acoustic technology barriers.** Preliminary research shows that bubbles and noises can frighten away the carp, thereby providing an important line of defense.
- **Third, immediately complete barriers to guard against carp entering the Great Lakes when floods occur.** While barriers are effective in preventing carp movements during normal water levels, floods can sweep carp over and around those barriers, as we have learned in the Mississippi River. Additional structures should be built specifically for flood conditions.
- **Fourth, conduct tagged fish research to validate the effectiveness of all primary and secondary barriers.** As government agencies

take action to construct barriers against carp movement, the effectiveness of those barriers should be assessed by tagging fish and tracking their movements.

- **Fifth, employ consistent measures to catch fish, such as electro-fishing, netting and commercial fishing.** These actions will both reveal where the carp are located, as well as reduce their populations, while providing a minimal interruption to navigation.
- **Sixth, fund research on Asian carp-specific biological control agents.** A federal agency, such as the U.S. Geological Survey, should be immediately tasked with identifying and developing agents that will either kill the carp or inhibit their reproduction, as has been done successfully with other invasive species.
- **Seventh, sample barges and other vessels for Asian carp or their eggs.** AWO supports efforts to determine all possible ways that the carp could travel into the Great Lakes. We are currently involved with a Coast Guard-led working group that will make certain that the ballast tanks of barges are not inadvertently transporting juvenile carp or carp eggs.
- **Eighth, impose further restrictions on the importation of aquatic invasive species.** Asian carp were purposely brought into the Unites States to clean fish farms; however, floods swept them into the Mississippi river. AWO supports legislation that would impose significant penalties for knowingly bringing aquatic invasive species into the United States or further transporting the species beyond its current habitat.
- **Finally, conduct more scientific studies about the ability of carp to survive within the Great Lakes ecosystem.** It has not yet been proven that Asian carp can establish a self-sustaining population in the Great Lakes ecosystem. However, if studies prove that survival is possible, then immediate action must be taken to mitigate population growth.

AWO believes that there are a number of options for controlling the Asian carp without putting the Chicago-area economy, jobs, environment and quality of life in jeopardy. The environment and the economy are inexorably linked; any measures to alter one will invariably affect the other. It is crucial that we act quickly and in concert with government, industry and environmental stakeholders to make sound decisions that benefit both the environment and the economy in the Great Lakes and Western Rivers.

Proposals have been made in both legislation and litigation to close locks in the Chicago Waterway System. We strongly oppose those proposals. These ideas do not meet the test of our fundamental principle to ensure that appropriate actions balance environmental protection with commercial sustainability. Closing the locks would stop barge traffic, replacing the lower emissions and high capacity of barge transportation with crippling levels of truck and rail congestion, coupled with a commensurate increase in air pollution. Proposals to close locks also wrongly assume that those structures are watertight barriers to fish, which they are not. And, speaking personally, closing locks would in all likelihood cause Illinois Marine Towing to go out of business, leading to the loss of over 100 jobs for the shore-side and vessel personnel of our company. Our company, and the other vessel operators working on the Illinois waterways, provides family wage employment to hard-working Americans. Together with government agencies and other concerned stakeholders, we need to develop effective solutions to stopping the Asian carp in a way that does not sacrifice jobs at a time when they are in such short supply.

Conclusion

Madam Chairman, Mr. Ranking Member, and Members of the Subcommittee, thank you for the opportunity to testify today. This committee has a history leadership in finding solutions to complex and challenging public policy issues without framing them as an "either, or" decision. AWO is a committed to working with you, with federal and state agencies and with other stakeholders of the Great Lakes and Western Rivers to ensure a balanced approach to environmental stewardship and economic sustainability. We are convinced that both goals can be realized.

Thank you. I look forward to answering any questions you may have.

In: Troubled Waters: Asian Carp and the... ISBN: 978-1-61761-766-9
Editors: Deborah M. Maeda

Chapter 11

TESTIMONY OF JOEL BRAMMEIER, PRESIDENT AND CEO, ALLIANCE FOR THE GREAT LAKES, BEFORE THE SUBCOMMITTEE ON WATER RESOURCES AND ENVIRONMENT, HEARING ON "ASIAN CARP AND THE GREAT LAKES"

INTRODUCTION

Madam Chair, thank you for inviting me to appear before this subcommittee to discuss the urgent threat posed to the Great Lakes by Asian carp. My name is Joel Brammeier, and I am the president and CEO of the Alliance for the Great Lakes. Formed in 1970, the Alliance is the oldest independent citizens' Great Lakes protection organization in North America. I am also a member of the steering committee of the Healing Our Waters – Great Lakes Coalition.

I and dozens of other dedicated citizens and experts have worked for many years in an advisory capacity to the U.S. Army Corps of Engineers and Illinois Department of Natural Resources to support the protection of the Great Lakes from invasion by bighead and silver carp. As a member of the Dispersal Barrier Advisory Panel,[1] I have offered my voice for nearly the last decade to a team of professionals that has provided sound advice, helpful critique and robust discussion among businesses, non-governmental organizations and agencies at all levels of government with a single task at hand: to stop Asian carp from establishing in the Great Lakes.

We can accomplish that task. But only if every decision we make today, tomorrow and next week is dedicated to one purpose: the permanent prevention of new invasive species movement into the Great Lakes via the Chicago Waterway System (CWS).

Creating an Artificial Connection

Behind nearly every invasive species in the Great Lakes are the hands of human intervention, and Asian carp are no different. The Great Lakes are connected to multiple other saline and fresh water bodies via artificial canals constructed during the 19th and 20th century. The Chicago Sanitary and Ship Canal (CSSC) and Portage Canal in Wisconsin connect Lake Michigan to the Illinois River, while multiple canals in Ohio connect Lake Erie to the Ohio River. The various canals of the New York State Canal System (NYSCS) and the Welland Canal not only link Lake Erie to the Hudson River and Atlantic Ocean, but also provide a western route from Lake Ontario to Lake Erie that is otherwise naturally blocked by Niagara Falls. Zebra mussels and round gobies have spread from the Great Lakes to the Mississippi River via the CSSC, while blueback herring, white perch and sea lamprey have all likely invaded the upper Great Lakes via the NYSCS.

In Chicago, the motivation for creating the artificial connection to the Mississippi was protection of the city's drinking water. Between 1860 and 1900, the North and South Branches of the Chicago River became a focus of industrial activity, including meat packing, slaughterhouses, distilleries and lumber mills. As the city grew rapidly, untreated sewage from homes and industries flowed to the rivers and into Lake Michigan, the primary source of drinking water for Chicago. The rivers became open sewers hosting bacteria and viruses causing typhoid, cholera, dysentery and other waterborne diseases.

In response, the Illinois General Assembly adopted the Sanitary District of Chicago Enabling Act in 1889. The legislation led to the creation of the Sanitary District of Chicago, the predecessor of the Metropolitan Water Reclamation District of Greater Chicago (MWRDGC). Soon after, its board of trustees, subscribing to the popular 19th century belief that "dilution is the solution to pollution," reversed the flows of the North and South Branches of the Chicago Rivers and the Calumet River away from Lake Michigan, and diverted clean lake and polluted river water downstream toward the Mississippi River.

By 1900, the man-made CSSC connected the South Branch of the Chicago River to the Des Plaines River. The artificial North Shore and Calumet-Sag Channels were completed in 1910 and 1922, respectively. Chicago's raw sewage, industrial wastes, and urban storm water were now directed away from the Great Lakes into the Des Plaines, Illinois, and Mississippi Rivers, with the unintended consequence of binding the ecology of the Great Lakes and Mississippi River watersheds and leading a potentially devastating fish across the threshold of Lake Michigan.

THE CHICAGO WATERWAY SYSTEM

The highly engineered combination of natural rivers and artificial canals continues to serve its intended 19th century purpose of disposing of wastewater, but 120 years of use have added layers of complexity. Many of the natural portions of the system have been channelized with attendant loss of natural habitats. The state of Illinois is authorized by the U.S. Supreme Court to stream up to 2.1 billion gallons of Great Lakes water per day – much of it as treated wastewater - past the channel walls of steel and limestone. The CWS is home to a recreational boating network with 35 – 70,000 recreational vessels moving between Lake Michigan and the inland waterways each year. Dozens of boat operators make their living moving tourists between downtown Chicago and Lake Michigan. The system supports a steady, although not growing, annual traffic of 20-25 million tons of bulk commodities such as coal, petroleum products and construction materials.[2]

In short, the Chicago metropolitan area has built itself with pride on the backbone of a 19th century engineering marvel bridging two great watersheds. This connection between the waterways and Lake Michigan was once seen as critical to opening the continent to trade and to feed in cool, clean Lake Michigan water to keep the rivers from reverting to open sewers. But the stark reality that this complex system created an aquatic superhighway for the jumbo-sized Asian carp and other invasive species to travel between Lake Michigan and Mississippi watersheds calls the question of whether this connection is as critical today as it seemed 10, 50 or 120 years ago.

THE LEVEL OF THREAT

When deciding whether to permit water pollution, regulators consider the ability of the environment to absorb the pollution against the potential impact on the health of humans, fish and wildlife. Conventional pollution permits take a "damage control" approach; there are dozens of pollutants that the Great Lakes tolerate in small amounts. Other more pernicious contaminants, such as mercury, are targeted for reduction and even elimination.

For some pollutants, however, the rules of damage control do not hold. Zero tolerance is the only appropriate response to the biological pollution of invasive species. There is no diluting their impacts to some unnoticed background level. By definition, these organisms establish in an ecosystem by outmuscling native species for food and reproductive opportunities. Once established, the chances of successful control are minimal.

Great Lakes fishery experts before this subcommittee today can speak thoroughly to the threat posed to the Great Lakes by Asian carp. I will summarize my perspective with three simple facts:

- In the Mississippi River basin where low-value Asian carp have invaded, they have established in great numbers and outcompeted native fish (Chapman 2003).
- U.S. (Kolar et al 2005) and Canadian (Mandrak and Cudmore 2004) risk assessments indicate that the Great Lakes have multiple carp-friendly habitats, including Green Bay, west Michigan, Saginaw Bay, Lake St. Clair, and western Lake Erie. These are some of the most popular boating and fishing spots in the region.
- The industries most at risk from an Asian carp invasion – sportfishing and recreational boating – contribute $7 billion and $16 billion (Great Lakes Commission 2007) annually, respectively, to the economies of the Great Lakes states.

Compounding these risks is the critical fact that, based on expert judgment that the eDNA monitoring method indicates live fish presence (Lodge 2009), carp are very likely already in the waters of Lake Michigan. This evidence demands that federal agencies immediately set a goal of beating back carp populations to zero as far from Lake Michigan as possible before they are able to establish in the Great Lakes

ECOLOGICAL SEPARATION

The presence of Asian carp in the Great Lakes and knowledge of the impacts of past invasions creates a strong incentive to act now to permanently protect both the Great Lakes and Mississippi River. Even if existing electrical barriers in the CSSC operate as designed, they will not last forever, nor will they ever achieve guaranteed 100 percent effectiveness. With the passage of time – through human error, an accident, wear and tear, or a natural disaster – the effectiveness of the barriers will be compromised.

The permanent solution is not reliance on technology, but on what we call "ecological separation," or no movement of live organisms between the Great Lakes and the Mississippi River via the canals – up to and including permanent physical barriers in the CWS. At first blush, in face of billions of gallons of water and millions of tons of cargo, this simple idea seems audacious.

However, a close look at water flows and navigation patterns suggests otherwise. Sewage treatment upgrades have diminished the need to maintain a connection to Lake Michigan water to maintain river water quality; the volume of these "discretionary diversions" has trended down for the last 20 years and will continue to decrease. Likewise, the need to allow combined sewage to flow into Lake Michigan for flood control has dramatically declined and will continue to do so with the completion of the local "Deep Tunnel" project. Commodity deliveries and loading via barge are confined in bulk to the Chicago Sanitary and Ship Canal and lower portions of the Calumet River, with comparatively little traffic moving through downtown Chicago or into the open waters of Lake Michigan. Barely more than 1,000 recreational boats move through Lockport lock annually. We can no longer afford to assume that 71 miles of canal in a century-old layout are required to get the job done, particularly when the job creates a massive external liability for the Great Lakes.

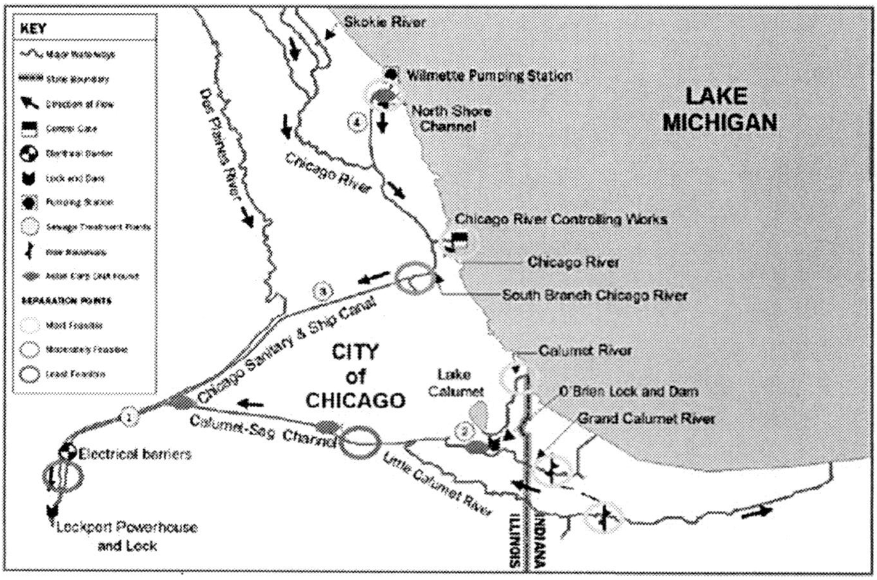

Figure 1. Potential locations of a physical separation of the Great Lakes and Mississippi River, 1/15/2010.

In light of these findings, a feasible separation scenario will:

- Stop all life stages of aquatic organisms moving between the Great Lakes and Mississippi River via the CWS;
- Accommodate the vast majority of commodity traffic within the waterways, as only a fraction of that traffic enters Lake Michigan;
- Provide new methods of moving recreational boaters between the Illinois River and Lake Michigan;
- Anticipate very occasional overflows into Lake Michigan to prevent local flooding, while improving the water quality of the lake and waterways; and
- Serve as a one-time payment for 100 percent effective, permanent protection against invasives.

Separation is not a new concept. In 2003, a local, state and federal interagency group hosted the Chicago Aquatic Invasive Species Summit for the express purpose of setting an agenda "beyond the barriers." This conference of more than 70 experts from around the region and the world agreed that stopping water movement between the Great Lakes and

Mississippi River was the only way to interrupt the stream of live organisms (City of Chicago 2005).

Two key steps were taken to follow up on this recommendation. In recognition that the separation effort was at the beginning of a long road of research and evaluation, the Great Lakes Fishery Commission and Great Lakes Fishery Trust supported an initial "prefeasibility" analysis of separation. This work, authored by a team including myself, Scudder Mackey (University of Windsor/Habitat Solutions, Inc.) and Irwin Polls (Ecological Monitoring and Assessment) identified several locations on the CWS that should be targeted for a full feasibility analysis (Figure 1). Our research reinforced the need to pursue a policy of ecosystem separation: in a series of interviews with stakeholders from around the Great Lakes in the U.S. and Canada, including decision-makers at all levels of government, it became clear that a goal of 100% effectiveness in stopping species movement was critical to the effort.

Stakeholders also recognized that the electrical barriers alone were unlikely to provide the permanent protection desired for the lakes. Knowing that any changes to the waterway would require significant federal investment and oversight, the Alliance and many others sought the necessary authorization for the U.S. Army Corps of Engineers to perform a Great Lakes and Mississippi River Interbasin Transfer Study, which became part of the *Water Resources Development Act of 2007*. While we are encouraged that the Corps has committed to developing an environmental impact statement for this work, we are concerned that few advance steps have been taken to tap the substantial expertise on the issue outside of federal agencies nearly two years after the authorization passed. We cannot afford to miss an opportunity to bring the best minds to bear on solving this problem.

RECOMMENDATIONS

We do not yet have a single permanent solution in hand. Each choice we make today is an opportunity to insure the Great Lakes against an Asian carp invasion. The smarter the choice, the longer the insurance policy extends and the more time we have to find and build a permanent solution. The policy and management actions below[3] are our best opportunity to keep the few carp already in Lake Michigan from becoming a permanent problem for all of the lakes:

Policy Needs

1. Federal and state agencies must immediately describe a short-term contingency plan with clear and firm triggers for action. All actions should flow from a singular goal of beating back existing populations of carp to zero as far from Lake Michigan as possible.
2. The U.S. Army Corps of Engineers must expedite the Chicago portion, including all National Environmental Policy Act (NEPA) requirements, of the Great Lakes and Mississippi River Interbasin Transfer Study so that it is completed by September 30, 2011 instead of 2014 as is currently expected. Implementation should begin in fiscal year 2012.
3. Provide clear agency authority to implement emergency actions through a public Memorandum of Understanding between agencies or some other public mechanism, with U.S. EPA in a lead role given its leadership in coordinating government action through the Great Lakes Interagency Task Force.
4. The Corps must complete the Dispersal Barrier Efficacy Study by August 2010. Immediately begin implementation of the flood control measures recommended in the Study's first interim report for completion by fall 2010.

Management Needs

1. Locks and sluice gates leading to Lake Michigan must be operated and managed in a way that reduces further transfer of Asian carp into Lake Michigan, including temporary closure if appropriate. Temporary changes in lock operations can slow movement of additional Asian carp toward Lake Michigan. Contingency plans should address needs for flood control and emergency response.
2. Operate the Dispersal Barrier System at optimal power and frequency and expedite both the completion of Barrier IIB and upgrade of Barrier I by the end of summer 2010.
3. Immediately install interim barriers in the Grand Calumet and Little Calumet Rivers, as necessary, to prevent Asian carp from migrating to Lake Michigan through Indiana.
4. Eliminate any risk of Asian carp bypassing the Dispersal Barrier System by waterway traffic, including the strict enforcement of the

Coast Guard's prohibition of ballast and bilge water transfers through the dispersal barriers.

5. Expand eDNA testing and consider eDNA results as actionable indicators of live Asian carp presence.

I encourage this subcommittee to work with the administration to ensure the fulfillment of these needs as soon as possible.

CONCLUSION

The false choice between Great Lakes protection and a freeflowing economy has taken center stage in the public debate about the urgent need to protect the Great Lakes from Asian carp invasion. Running a close second is the contention that threat to property, life and limb by flooding and disease is a precondition of changing how we use the Chicago Waterway System. This does a disservice to the reality on the ground: that the Great Lakes and Mississippi River are being held hostage by outmoded infrastructure and dated assumptions that how business gets done is the way business has to be done.

A rare opportunity stands before two great waters of the United States today. We understand the damage that has already occurred. We can predict the irreparable harm to the Great Lakes if we fail. We have the tools and knowledge in hand to stop the problem before it starts. We cannot accept the outdated zero-sum thinking of industry or environment, clean water or invasive species. The original engineering feat of the Chicago Waterway System protected the Great Lakes, but transferred costs to others – costs that were not entirely apparent in 1890 but are a hole in the wallet today. Those with the vision to see and will to transform understand that this backbone of the largest Great Lakes city must either stretch and strengthen with time, or collapse under its own weight. I look forward to working with this subcommittee and everyone engaged on this matter to create a sustainable legacy for the CWS that outlasts both me and the original project. Madam Chair, thank you for holding this hearing and I look forward to assisting on any actions this subcommittee can take to support this effort.

REFERENCES

Brammeier, J. B. & Scudder Mackey and Irwin Polls. (2008). Preliminary feasibility of ecological separation of the Mississippi River and the Great Lakes to prevent the transfer of aquatic invasive species. Online at http://www.greatlakes.org/Document.Doc?id=473.

Chapman, D. (2003). Bighead and silver carp in the Mississippi and Missouri Rivers. Online at infolink.cr.usgs.gov/Science/Documents/invasive_c arp.pdf.

City of Chicago. (2005). Closing the revolving door: summary of the aquatic invasive species summit proceedings. Online at *http://egov.cityof chicago.org/w*ebportal/COCWebPortal/COC_ATTACH/Aquatic_Invasive _Speci es_Summary.pdf.

Great Lakes Commission. (2007). Great Lakes recreational boating's economic punch. Online at http://glc.org/recboat/pdf/rec-boating-final-small.pdf.

Kolar, C. S., et al. (2005). Asian carps of the genus Hypophthalmichthys (pisces, cyprinidae)--A biological synopsis and environmental risk assessment. Online at www.fws.gov/contaminants/OtherDocuments/ ACBSRAFinalReport2005.pdf.

Lodge, David. (2009). Declaration of David M. Lodge in Appendix to memorandum for the United States in Opposition, Supreme Court Nos. 1, 2 and 3, Original. Online at *http://www.supremecourtus.gov/Sp ecMastRpt/A*pptoMemfortheUSinOpposition.pdf

Mandrak, N. E. & Cudmore, B. (2004). Risk assessment for Asian carps in Canada. Online at *www.dfo-mpo.gc.ca/csas/Csas/DocREC/2004* /RES2004_103_E.pdf.

End Notes

[1] http://seagrant.wisc.edu/AIS/Default.aspx?tabid=393.

[2] All data on navigation are published by the U.S. Army Corps of Engineers Waterborne Commerce Statistics Center. Data were extracted and organized from Corps databases via a proprietary program written by Scudder Mackey of Habitat Solutions, Inc. and are available upon request. Original databases are available for public download at http://www.iwr.usace.army.mil/ndc/wcsc/wcsc.htm.

[3] Adapted from recommendations developed by the Healing Our Waters-Great Lakes Coalition, available from Chad Lord, Water Program Director, National Parks Conservation Association.

In: Troubled Waters: Asian Carp and the… ISBN: 978-1-61761-766-9
Editors: Deborah M. Maeda

Chapter 12

CHICAGO SANITARY AND SHIP CANAL DISPERSAL BARRIER SYSTEM

United States Army Corps of Engineers

Location and Purpose: The dispersal barriers are located in the Chicago Sanitary and Ship Canal (CSSC), which is a man-made waterway creating the only continuous connection between Lake Michigan and the Mississippi River basin. The dispersal barrier system was developed to prevent the spread of invasive fish species between these watersheds.

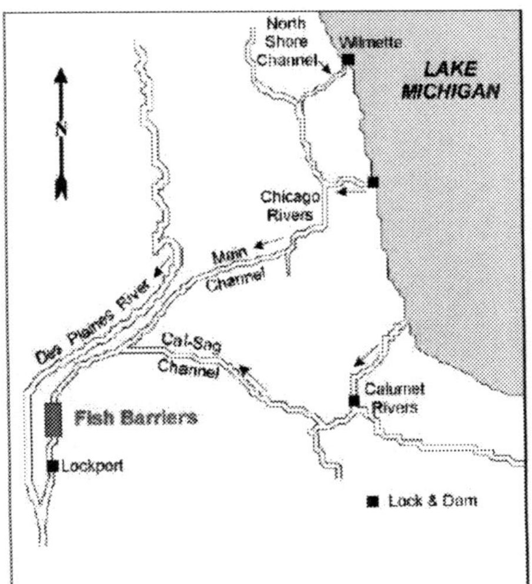

Project History: Operation of the first barrier (Barrier I) began in April 2002 by the Corps, demonstrating a new technology for preventing the spread of aquatic nuisance species.

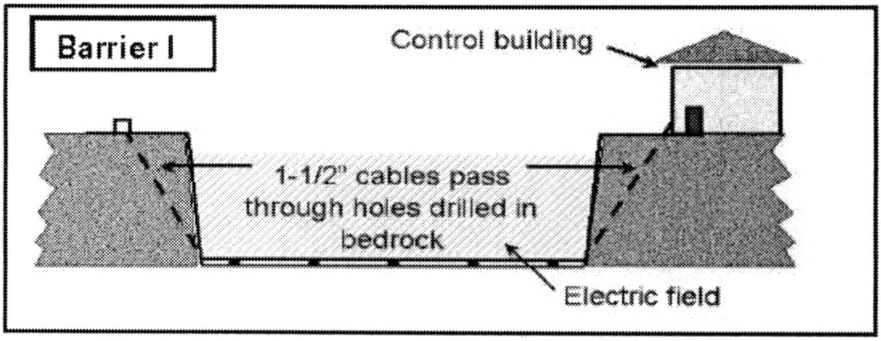

Barrier I, which is located at river mile 296.5 in Romeoville, IL, is formed of steel cables (see diagram below right) that are secured to the bottom of the canal. A low-voltage, pulsing DC current is sent through the cables, creating an electric field in the water. The electric field is uncomfortable for fish and they do not swim across it.

In 2004, the Corps initiated construction of a permanent barrier (Barrier II) to prevent the migration of fish, including Asian carp, between the watersheds. Barrier II, which is located 800 to 1,300 feet downstream of Barrier I, also uses a pulsed electric field, but includes several design improvements identified during monitoring and testing of Barrier I.

Barrier II is able to generate a more powerful electric field over a larger area and consists of two sets of electrical arrays and control houses, known as Barriers IIA and IIB. Each control house and set of arrays can be operated independently, but ultimately goal is to operate both concurrently.

In 2007, Congress authorized the Corps to complete Barrier II, to upgrade Barrier I and make it permanent, and to operate the barrier system at full federal cost.

Status: Barrier I and Barrier IIA are operating continuously. Barrier IIB is partially constructed.

Due to its original demonstration status, Barrier I was designed and built with materials that were not intended for long-term use. Significant repairs were successfully completed at Barrier I in October 2008. These repairs will allow Barrier I to remain in service for several more years until Barrier IIB is completed and fully functional. Once Barrier II is fully operational, Barrier I will be taken off line and replaced with a more permanent facility.

Barrier IIA was activated in April 2009 at the same operational settings as Barrier I. Construction of Barrier IIB began in the Fall of 2009 and will be completed in 2010.

The Corps also directs a monitoring program to identify the location of Asian carp relative to the barriers. In the summer of 2009 the Corps contracted with the University of Notre Dame to deploy environmental DNA monitoring (eDNA), a new monitoring method developed at Notre Dame. This method does not rely on direct observation of Asian carp to evaluate their presence. The eDNA method has detected Silver and Bighead carp closer to the barriers than previously believed. The closest eDNA detection of Asian carp in the CSSC is approximately one mile downstream of the barriers. No positive detections in the CSSC have occurred upstream of the barriers.

A study by independent researchers indicated that operating parameters used at Barrier I may not be effective for deterring smaller fish so the Corps initiated a research program to identify the optimal operating settings for the dispersal barriers. Based on initial results from this research the operating settings at Barrier IIA were increased in August 2009 to an operational setting of 15 pulses per second with each pulse 6.5 milliseconds long and maximum in-water field strength of 2 Volts per inch to repelling both adult and juvenile fish. The operating settings at Barrier I were not adjusted because the equipment at Barrier I is unable to operate at the higher operating settings.

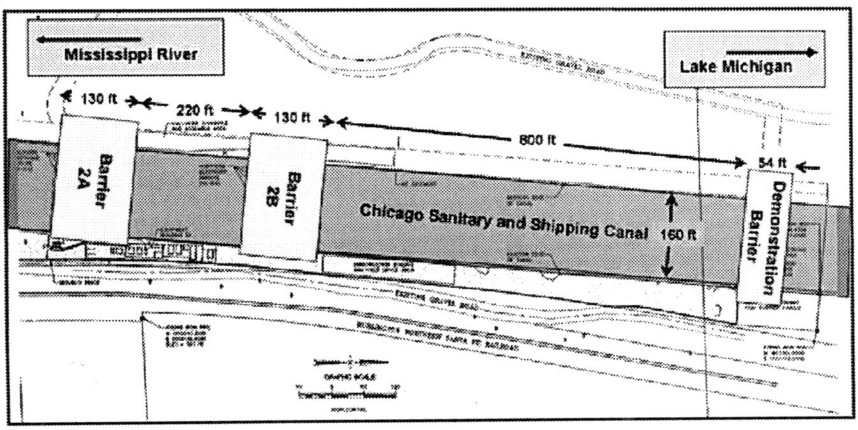

Studies: Two studies related to the CSSC Barriers were authorized by Congress in 2007. The first is an investigation of hazards that might compromise the effectiveness of the barriers, including potential bypassing of the barriers through the Des Plaines River or other waterways during flood flows. This study is part of the ongoing barriers project and was initiated in 2009. The second study is a comprehensive investigation of the feasibility of other approaches to prevent the inter-basin transfer of aquatic nuisance species between the Great Lakes and Mississippi River basins. The feasibility study is a separate project that was also initiated in 2009.

Maintenance: Performing scheduled maintenance is necessary to maintain reliability of the structures and minimize the risk of unplanned outages. During the first week of December 2009, the U.S. Army Corps of Engineers is planning to perform scheduled maintenance on Barrier IIA. During the maintenance shutdown, Barrier I will remain active. However, because of late summer detection of Asian carp near the barrier system and concern that Barrier I may not be effective in deterring juvenile fish, a fish toxicant called rotenone will be applied to the canal between the barrier and the Lockport Lock and Dam. This will allow for the removal of Asian carp and other fish to keep them from advancing past the barrier toward Lake Michigan. The Illinois Environmental Protection Agency will conduct water sampling to ensure that the waters of the state are protected.

Project Manager: Chuck Shea, USACE Chicago District, 312-846-5568.

Chapter Sources

The following chapters have been previously published:

Chapter 1 – This is an edited, excerpted and augmented edition of a United States Congressional Research Service publication, Report Order Code R41082, dated June 25, 2010.

Chapter 2 – This is an edited, excerpted and augmented edition of a United States House of Representatives Subcomittee on Water Resources and Environment Staff Hearing, dated February 5, 2010.

Chapter 3 – These remarks were delivered as Statement of Cameron Davis, before the House of Representative Subcommittee on Water Resources and the Environment, dated February 9, 2010.

Chapter 4 – These remarks were delivered as Statement of Major General John Peabody, before the House of Representative Subcommittee on Water Resources and the Environment, dated February 9, 2010.

Chapter 5 – These remarks were delivered as Statement of Marc Miller, before the House of Representative Subcommittee on Water Resources and the Environment, dated February 9, 2010.

Chapter 6 – These remarks were delivered as Statement of Rebecca Humphries, before the House of Representative Subcommittee on Water Resources and the Environment, dated February 9, 2010.

Chapter 7 – These remarks were delivered as Statement of Matt Frank, before the House of Representative Subcommittee on Water Resources and the Environment, dated February 9, 2010.

Chapter 8 – These remarks were delivered as Statement of David M. Lodge, before the House of Representative Subcommittee on Water Resources and the Environment, dated February 9, 2010.

Chapter 9 – These remarks were delivered as Statement of Dr. Michael J. Hansen, before the House of Representative Subcommittee on Water Resources and the Environment, dated February 9, 2010.

Chapter 10 – These remarks were delivered as Statement of Del Wilkins, before the House of Representative Subcommittee on Water Resources and the Environment, dated February 9, 2010.

Chapter 11 - These remarks were delivered as Statement of Joel Brammeier, before the House of Representative Subcommittee on Water Resources and the Environment, dated February 9, 2010.

Chapter 12 - This is an edited, excerpted and augmented edition of a United States Army Corps on Engineers publication, dated November 13, 2009.

INDEX

O

P

Q

R

S